THE PRION PROTEIN

Edited by:

Jörg Tatzelt

Ludwig-Maximilians-University Munich
Germany

Savanna Press ... **s.com**

www.savannapress.com

British Library Cataloguing-in-Publication Data

A catalogue record for this book is available from the British Library

ISBN: 978-0-9543335-2-2

Description or mention of instrumentation, software, or other products in this book does not imply endorsement by the author or publisher. The author and publisher do not assume responsibility for the validity of any products or procedures mentioned or described in this book or for the consequences of their use.

Contents

Contributors

Yasmine Aguib
Institute of Virology
Prion Research Group
Technische Universität München
Trogerstr. 30
81675 Munich
Germany

Flavio H. Beraldo
Ludwig Institute for Cancer Research
Hospital Alemão Oswaldo Cruz
São Paulo SP
Brazil

Dola Das
The Department of Pathology
Case Western Reserve University
2103 Cornell Road
Cleveland
Ohio 44106
USA

Glaucia N. Hajj
Ludwig Institute for Cancer Research
Hospital Alemão Oswaldo Cruz
São Paulo SP
Brazil

David A. Harris
Department of Cell Biology and Physiology
Washington University School of Medicine
St. Louis
MO 63110 USA
Current address:
Department of Biochemistry
Boston University School of Medicine
72 East Concord Street
Boston
MA 02118
USA

Andreas Heiseke
Institute of Virology
Prion Research Group
Technische Universität München
Trogerstr. 30
81675 Munich
Germany

Kil Sun Lee
Associação Fundo de Incentivo à Psicofarmacologia
São Paulo SP
Brazil

Marilene H. Lopes
Ludwig Institute for Cancer Research
Hospital Alemão Oswaldo Cruz
São Paulo SP
Brazil

Rafael Linden
Instituto de Biofísica Carlos Chagas Filho
Universidade Federal do Rio de Janeiro
Rio de Janeiro RJ
Brazil

Giovanna R. Mallucci
MRC Toxicology Unit
Hodgkin Building
University of Leicester
Lancaster Road
Leicester LE1 9EH
UK

Vilma R. Martins
Ludwig Institute for Cancer Research
Hospital Alemão Oswaldo Cruz
São Paulo SP
Brazil

Margit Miesbauer
German Center for Neurodegenerative Diseases (DZNE)
Munich & Adolf-Butenandt-Institute
Neurobiochemistry
Ludwig-Maximilians-University Munich
Schillerstrasse 44
D-80336 München
Germany

Maradumane L. Mohan
The Department of Pathology
Case Western Reserve University
2103 Cornell Road
Cleveland
Ohio 44106
USA

Marco A. Prado
Robarts Research Institute
Departments of Anatomy and Cell Biology/Physicology
and Pharmacology
University of Western Ontario
London
Ontario
Canada
and
Department of Pharmacology
Federal University of Minas Gerais MG
Brazil

Helois E. Radford
MRC Toxicology Unit
Hodgkin Building
University of Leicester
Lancaster Road
Leicester LE1 9EH
UK

Angelika S. Rambold
German Center for Neurodegenerative Diseases (DZNE)
Munich & Adolf-Butenandt-Institute
Neurobiochemistry
Ludwig-Maximilians-University Munich
Schillerstrasse 44
D-80336 München
Germany
Current address:
Cell Biology and Metabolism Program
National Institute of Child Health and Human Development
National Institutes of Health
Bethesda
Maryland
USA

Hermann M. Schatzl
Institute of Virology
Prion Research Group
Technische Universität München
Trogerstr. 30
81675 Munich
Germany

Jessie A. Schepker
Department of Cell Biology and Physiology
Washington University School of Medicine
St. Louis
MO 63110
USA

Ajay Singh
The Department of Pathology
Case Western Reserve University
2103 Cornell Road
Cleveland
Ohio 44106
USA

Neena Singh
The Department of Pathology
Case Western Reserve University
2103 Cornell Road
Cleveland
Ohio 44106
USA

Isaac H. Solomon
Department of Cell Biology and Physiology
Washington University School of Medicine
St. Louis
MO 63110
USA

Jörg Tatzelt
German Center for Neurodegenerative Diseases (DZNE)
Munich & Adolf-Butenandt-Institute
Neurobiochemistry
Ludwig-Maximilians-University Munich
Schillerstrasse 44
D-80336 München
Germany

Konstanze F. Winklhofer
German Center for Neurodegenerative Diseases (DZNE)
Munich & Adolf-Butenandt-Institute
Neurobiochemistry
Ludwig-Maximilians-University Munich
Schillerstrasse 44
D-80336 München
Germany

Preface

The Prion Protein: Friend and Foe

Jörg Tatzelt

(PrP^{C}) into an aberrantly folded isoform designated scrapie prion protein (PrP^{Sc}) is the hallmark of a variety of neurodegenerative disorders collectively called prion diseases. They include Creutzfeldt-Jakob disease and Gerstmann-Stäussler-Scheinker syndrome in humans, scrapie in sheep, bovine spongiform encephalopathy (BSE) in cattle and chronic wasting disease (CWD) in free-ranging deer. In contrast to the deadly properties of misfolded PrP, PrP^{C} seems to possess a neuroprotective activity. Moreover, animal models indicated that the stress-protective activity of PrP^{C} and the neurotoxic effects of PrP^{Sc} are somehow interconnected.

In this book we have assembled a set of reviews from some leading scientists in the field to highlight the apparently incongruous activities of different PrP conformers. The chapters outline current research on celluar pathways implicated in the formation and signaling of neurotoxic and physiological PrP isoforms and delineate future research direction.

Vilma Martins comments on our current understanding about the physiologcial activity of PrP^{C} and its possible role as a neurotrophic factor. The observation that aberrant PrP conformers can cause neurodegeneration in the absence of infectious prion propagation is an expanding field highlighted by David Harris. Giovanna Mallucci expands on this topic and reports on the intriguing finding that the GPI anchor of PrP^{C} is required to mediate neurotoxic effects of scrapie prions.To follow up this track, Konstanze Winklhofer and I specifically concentrate on pathways implicated in the formation and neurotoxic properties of cytosolically localized PrP. Neena Singh illustrates that metal ions can have a significant impact on the processing of PrP. Finally, Hermann Schätzl provides insight into the role of autophagy in the propagation and clearance of PrP^{Sc}.

I am grateful to all contributing authors and editorial staff for their time and effort.

Chapter 1

Prion Neurotoxicity: Insights from Prion Protein Mutants

Isaac H. Solomon[1], Jessie A. Schepker[1] and David A. Harris

Abstract
The chemical nature of prions and the mechanism by which they propagate are now reasonably well understood. In contrast, much less is known about the identity of the toxic prion protein (PrP) species that are responsible for neuronal death, and the cellular pathways that these forms activate. In addition, the normal, physiological function of cellular PrP (PrP^{C}) has remained mysterious, hampering efforts to determine whether loss or alteration of this function contributes to the disease phenotype. Considerable evidence now suggests that aggregation, toxicity, and infectivity are distinct properties of PrP that do no necessarily coincide. In this review, we will discuss several mutant forms of PrP that produce spontaneous neurodegeneration in humans and/or transgenic mice without the formation of infectious PrP^{Sc}. These include an octapeptide insertional mutation, point mutations that favor synthesis of transmembrane forms of PrP, and deletions encompassing the central domain whose neurotoxicity is antagonized by the presence of wild-type PrP. By isolating the neurotoxic effects of PrP from the formation of infectious prions, these mutants have provided important insights into possible pathogenic mechanisms. These studies suggest that prion neurotoxicity may involve subversion of a cytoprotective activity of PrP^{C} via altered signaling events at the plasma membrane.

Introduction
Prion diseases are fatal neurodegenerative illnesses of man and animals. This group of disorders includes Creutzfeldt-Jakob Disease (CJD), kuru, fatal familial insomnia (FFI), Gerstmann-Sträussler syndrome (GSS), and new variant CJD in humans, as well as bovine spongiform encephalopathy in cattle, chronic wasting disease in deer and elk, and scrapie in sheep and goats (Prusiner, 2004). Patients affected with these disorders suffer from dementia and ataxia, and often display spongiform degeneration and amyloid deposition in their brains.

A wealth of evidence suggests that the central molecular event in prion diseases is the conformational conversion of PrP^{C}, a normal cell-surface glycoprotein, into PrP^{Sc}, an abnormal isoform that is infectious in the absence of nucleic acid (Aguzzi et al., 2008; Prusiner, 1998). The precise structural differences between the two PrP isoforms remain to be defined, although it is clear that PrP^{Sc} contains significantly more β-sheet and is more protease-resistant and aggregated than PrP^{C}. The conversion of PrP^{C} to PrP^{Sc} is thought to involve a templating mechanism in which the two forms physically interact.

Although we now have a detailed understanding of how prions propagate, the cellular mechanisms by which they kill neurons, and the toxic forms of PrP responsible, are poorly understood (Chiesa and Harris, 2001; Harris and True, 2006). Important insights into this issue have been obtained by analysis of PrP molecules carrying neurotoxic mutations. Several kinds of mutant PrP molecules induce spontaneous neurological illness in human beings or transgenic mice in the absence of infection from exogenous sources (Table 1 and Fig. 1). One category of such molecules are those carrying point or insertional mutations linked to human familial prion diseases. These mutants generally display PrP^{Sc}-like biochemical properties, and at least part of their pathogenicity is likely to depend on the toxic properties of the oligomeric protein aggregates that they form. A second category of mutations are those in the N-terminal signal sequence and hydrophobic domain that influence the membrane topology of PrP. A third category includes a series of deletion mutations encompassing the central region of PrP that endow the protein with a powerful neurotoxic activity suppressible by co-expression of wild-type PrP. Mutants in the last two categories are not aggregated or protease-resistant, and their effects are likely due to alterations in a physiological activity of PrP^{C}.

Importantly, none of these three categories of mutations is accompanied by the formation of infectious PrP^{Sc}. By isolating the neurotoxic effects of PrP from the propagation of infectious prions, these mutants have made it possible to focus on pathogenic mechanisms underlying the disease process. Some of these mechanisms turn out to be surprisingly similar to those associated with non-infectious neurodegenerative disorders such as Alzheimer's disease.

In this article, we will discuss work from our laboratory utilizing each of these three categories of PrP mutants. As a prelude, we will first review what is currently known about the PrP forms and cellular pathways underlying prion neurotoxicity, as well as our current understanding of the physiological function of PrP^{C}.

Prion neurotoxicity: what is the toxic molecule?
What form of PrP is responsible for killing neurons? It has commonly been assumed that PrP^{Sc} itself is the primary cause of neurodegeneration, based on the temporal and anatomical correlation between the accumulation of this form and the development of neuropathological changes.

[1]These two authors contributed equally to this work.

The Prion Protein 2010 Jörg Tatzelt (Ed.) Savanna Press, UK
ISBN: 978-0-9543335-2-2. savannapress.com

Table 1. Properties of PrP^C, PrP^Sc, and neurotoxic mutants of PrP

Molecule	Protease-resistance/ aggregation	Cellular localization	Infectivity	Disease onset (days)[a]	Pathological targets	Effect of wild-type PrP (expression level)
PrP^{C}	-	PM/Rafts	-	N/A	None	N/A
PrP^{Sc}	+++	PM, endosomes	+	Depends on strain/host	Neurons, WM	Required
$PG14^{Spon}$ [1-3]	+	ER	-	240	CGN, dendrites, axons/myelin	No effect (1X)
$PG14^{RML}$ [1, 2]	+++	Unknown	+	240 (post-inoculation)	CGN	Not tested
^{Ctm}PrP (L9R/3AV) [4-6]	-	ER (CHO cells); Golgi (neurons)	-	85-172[b]	CGN, HPN	Exacerbation (0.5-1X)
PrPΔ32-121 [7]	-	Not tested	-	42-56[c]	CGN	Rescue (0.5X)
PrPΔ32-134 [7]	-	PM	-	21-35[c]	CGN, WM	Rescue (0.5X)
PrPΔ94-134 [8]	-	Rafts	-	14[c]	WM	Rescue (2-3X)
PrPΔ105-125 [9]	-	PM/Rafts	-	7[c]	CGN, WM	Rescue (5X)

Abbreviations: CHO, Chinese hamster ovary; CGN, cerebellar granule neurons; ER, endoplasmic reticulum; HPN, hippocampal neurons; N/A, not applicable; PM, plasma membrane; WM, white matter.
[a]Disease onset varies with transgene expression level. Data are for highest expressing lines (1-3X).
[b]For mice on the *Prn-p*$^{+/+}$ background.
[c]For mice on the *Prn-p*$^{0/0}$ background.
References: [1]Chiesa et al., 2003; [2]Biasini et al., 2008; [3]Jeffrey et al., 2009; [4]Stewart et al., 2005; [5]Stewart and Harris, 2005; [6]Stewart et al., 2001; [7]Shmerling et al., 1998; [8]Baumann et al., 2007; [9]Li et al., 2007b.

However, there are a number of situations where this correlation is weak or absent. In several kinds of transmission experiments, for example, significant pathology and/or clinical dysfunction develop with little accumulation of PrP^{Sc} (Flechsig et al., 2000; Lasmezas et al., 1997; Manson et al., 1999). In addition, some familial prion diseases are not transmissible, and are not accompanied by the accumulation of protease resistant PrP (Brown et al., 1994b; Piccardo et al., 2001; Tateishi and Kitamoto, 1995; Tateishi et al., 1990; Tateishi et al., 1996). On the other hand, there are subclinical infections in which there is abundant PrP^{Sc} but little symptomatology, for example after inoculation of hamster prions into mice (Hill et al., 2000; Race et al., 2001).

Taken together, these situations argue that PrP^{Sc}, the infectious form of PrP, may not be the proximate cause of neuronal dysfunction and degeneration in prion diseases. Several alternative forms of PrP, distinct from both PrP^{C} and PrP^{Sc}, have therefore been hypothesized to be the primary neurotoxic species (designated PrP^{toxic}). In subsequent sections, we will discuss two possible candidates for PrP^{toxic} ($PG14^{spon}$ and ^{Ctm}PrP).

Prion neurotoxicity: what are the cellular pathways?

How do PrP^{Sc} or other toxic forms of PrP induce neuronal death? There is now a growing body of evidence that PrP^{C}, in addition to serving as a precursor of PrP^{Sc}, acts as a signal transducer or mediator of the neurotoxic effects of PrP^{Sc} (Harris and True, 2006). Support for this concept derives from several experimental situations in which expression of membrane-anchored PrP^{C} in target neurons is essential for conferring sensitivity to PrP^{Sc}-induced neurodegeneration or toxicity. First, *Prn-p*$^{0/0}$ neurons are resistant to the pathologic effects of PrP^{Sc} supplied from grafted brain tissue (Brandner et al., 1996) or from nearby astrocytes (Mallucci et al., 2003). Second, scrapie-inoculated mice expressing a glycolipid anchor-negative version of PrP develop minimal brain pathology and neurological dysfunction despite the accumulation of substantial amounts of PrP^{Sc} amyloid (Chesebro et al., 2005). Although interpretation of this experiment is complicated by the relatively low expression levels of anchorless PrP in these mouse lines, one possible implication is that PrP must be membrane-anchored to efficiently mediate a toxic signal (Aguzzi, 2005). Third, *Prn-p*$^{0/0}$ neurons in culture have been found to be resistant to apoptosis induced by exposure to the synthetic peptide PrP106-126, which has been used as a model for PrP^{Sc} (Brown et al., 1994a). Finally, a recent study has shown that PrP^{C} exerts a protective activity against cellular stress, and that PrP^{Sc} abrogates this activity by activating stress-related signaling cascades (Rambold et al., 2008).

Combined with evidence that PrP^{C} normally serves as a neuroprotective molecule (see below), these results taken together suggest a "subversion-of-function" hypothesis to explain prion-induced pathology (Harris and True, 2006):

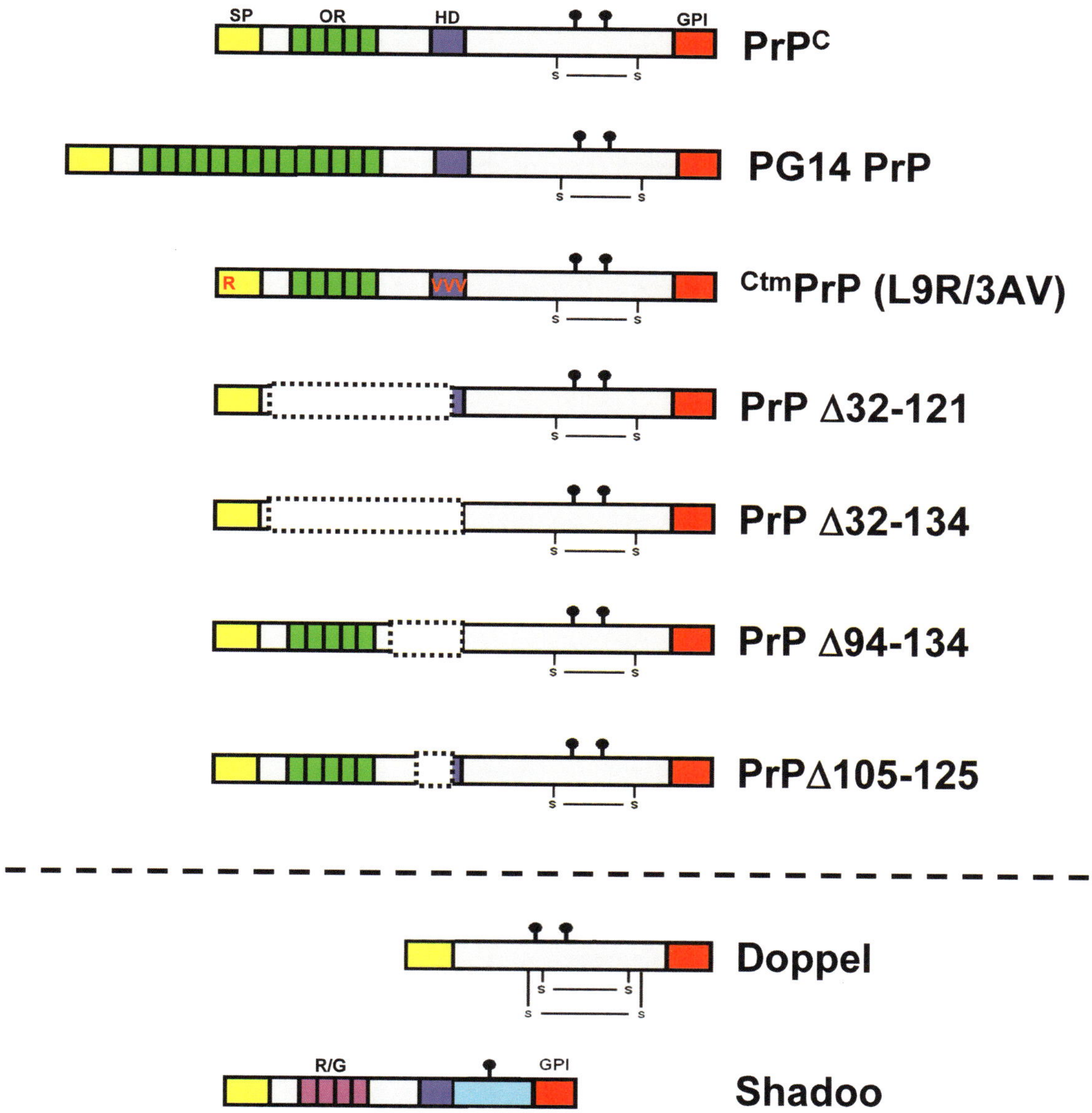

Figure 1. Schematic of wild-type and mutant PrP molecules, Doppel, and Shadoo. Structural domains are indicated by the colored blocks: SP (yellow), signal sequence; OR (green), octapeptide repeats; HD (blue), hydrophobic domain; GPI (red), glycosyl-phosphatidylinositol attachment signal; R/G (pink), arginine/glycine repeats of Sho. The lollipop symbols indicate sites of N-linked glycosylation, and the S—S symbols indicate disulfide linkages.

specifically, interaction with PrP^{Sc} (or other pathogenic intermediates) alters a normal, physiological activity of PrP^C in such a way that a neurotoxic stimulus is delivered. We will return to this mechanism later, when discussing the neurotoxicity of certain deleted forms of PrP.

Function of PrP^C

In order to specify how PrP^C function is altered or subverted in prion diseases, it is clearly essential to understand the normal, physiological role of the protein. Surprisingly, however, twenty-five years since PrP^C was first identified as an endogenous cellular protein, its physiological activity remains obscure (reviewed in Westergard et al., 2007). A variety of functions have been proposed, including roles in metal ion trafficking (Pauly and Harris, 1998), cell adhesion (Mange et al., 2002), and signal transduction (Mouillet-Richard et al., 2000). However, the evidence in favor of each of these hypothesized functions is not definitive.

Attempts to deduce the function of PrP^C from the phenotypes of PrP-null mice have been unrewarding. These mice display no major anatomical or developmental deficits, with the exception of lines in which the gene encoding Doppel (a PrP paralog) is artifactually up-regulated (Büeler et al., 1992; Manson et al., 1994). Subtle phenotypic abnormalities have been described in PrP knock-out mice at the cellular or behavioral level, but some of these have been controversial, and in any case they have not provided an unambiguous clue to the biological function of PrP^C (Steele et al., 2007). A recent report suggests a role for PrP^C in embryonic cell adhesion, based on the phenotype of zebrafish in which expression of a PrP homologue has been knocked down (Malaga-Trillo et al., 2009). Whether this purported function is conserved in mammalian PrP remains to be determined.

Several intriguing lines of evidence have emerged recently suggesting that PrP^C may exert a cytoprotective activity, particularly against stresses (either internal or environmental) that initiate an apoptotic program (reviewed in Roucou et al., 2004; Roucou and LeBlanc, 2005). First, PrP over-expression rescues cultured neurons, some mammalian cell lines and yeast cells from pro-apoptotic stimuli, including Bax expression, serum withdrawal, and cytokine treatment (Bounhar et al., 2001; Diarra-Mehrpour et al., 2004; Kuwahara et al., 1999; Li and Harris, 2005; Roucou et al., 2005). Second, there is evidence that endogenous PrP protects cells against oxidative and pathologic stressors. For example, neurons cultured from *Prn-p*$^{0/0}$ mice display several subtle abnormalities related to increased susceptibility to oxidative stress (Brown et al., 2002). Moreover, after ischemic or traumatic brain injury, lesion size is larger in *Prn-p*$^{0/0}$ compared to wild-type mice (Hoshino et al., 2003; McLennan et al., 2004; Sakurai-Yamashita et al., 2005; Spudich et al., 2005). Finally, *Prn-p*$^{0/0}$ mice are more susceptible to kainate-induced seizures (Rangel et al., 2007) and their retinal photoreceptors are more prone to light-induced degeneration (Frigg et al., 2006).

To explore the cytoprotective activity of PrP^C, we recently attempted to reproduce several of the published assays for this activity (Christensen and Harris, 2008). In one set of experiments, we found that PrP over-expression had a minimal effect on the death of MCF-7 breast carcinoma cells treated with TNF-α, and on *Prn-p*$^{0/0}$ immortalized hippocampal neurons (HpL3-4 cells) subjected to serum deprivation. In a second set of assays, we observed only a small difference in viability between cerebellar granule neurons cultured from PrP-null and control mice in response to activation of endogenous or exogenous Bax. Although our results do not rule out a cytoprotective activity of PrP^C, they suggest that existing cell culture systems may be inadequate for modeling this activity.

Mutant PrP molecules associated with inherited prion diseases

Studies of mutant PrP molecules associated with inherited prion diseases have provided important clues to the molecular characteristics of PrP^{toxic}, and the mechanisms by which it is pathogenic. Approximately 10% of the cases of CJD and all cases of GSS and FFI are linked to dominantly inherited, germline mutations in the PrP gene on chromosome 20 (Mead, 2006). The mutations are presumed to favor spontaneous conversion of the protein to the PrP^{Sc} state without the necessity for contact with exogenous infectious agent. Point mutations occur in the C-terminal half of the PrP molecule, and are associated with either CJD, GSS or FFI. Insertional mutations, which are associated with a variable phenotype that can include features of CJD or GSS, consist of one to nine additional copies of a peptide repeat that is normally present in five copies in the N-terminal half of the protein.

A number of studies have used transfected cell lines to analyze the biochemical and cell biological properties of PrP molecules carrying disease-associated mutations (reviewed by Harris, 2003). These studies have revealed that some, but not all mutants, display PrP^{Sc}-like biochemical properties including aggregation propensity as well as resistance to proteases and a GPI-specific phospholipase (Lehmann and Harris, 1996; Priola and Chesebro, 1998). Correlating with these abnormal biochemical properties, some mutants display altered subcellular localization, including partial retention in the ER (Ivanova et al., 2001), or retrotranslocation into the cytoplasm with subsequent degradation by the proteasome (Lorenz et al., 2002). Interestingly, some disease-associated PrP mutants are identical to wild-type PrP in terms of their biochemical properties and cellular distribution (Harris, 2003), raising the possibility that these molecules are pathogenic, not because they misfold and aggregate, but because the mutation subtly alters a physiological activity normally associated with PrP^C. Unfortunately, expression in transfected cells of PrP molecules carrying disease-associated mutations is generally not cytopathic, making it difficult to analyze the neurotoxic mechanisms underlying these mutants. For this reason, we and others have turned to transgenic models of familial prion diseases.

A transgenic model of an octapeptide insertion: Tg(PG14) mice

Tg(PG14) mice express the mouse homologue of a nine-octapeptide insertional PrP mutant associated with a familial prion disease of humans (Duchen et al., 1993; Krasemann et al., 1995; Owen et al., 1992) (Fig. 1). These animals display a spontaneous, progressive neurodegenerative disease with symptoms of ataxia, cerebellar granule cell loss, gliosis, and PrP deposition (Chiesa et al., 2000; Chiesa et al., 1998). Tg(PG14) mice spontaneously accumulate in their brains an insoluble and weakly protease-resistant form of the mutant protein (Chiesa et al., 2003). This form (designated $PG14^{Spon}$) is highly neurotoxic, but is not infectious in animal bioassays. In contrast, when Tg(PG14) mice are inoculated with the RML strain of prions, they accumulate a different form of PG14 PrP (designated $PG14^{RML}$) that is highly protease-resistant and infectious in animal transmission experiments (Chiesa et al., 2003). Consistent with other published work, these studies emphasize that infectivity and pathogenicity are two distinct properties of PrP.

To gain insight into the molecular determinants of infectivity and pathogenicity, we have undertaken a biochemical characterization the biochemical properties of $PG14^{Spon}$ and $PG14^{RML}$. Our initial studies (Chiesa et al., 2003) indicated that these two forms, although conformationally related, differed in their quaternary structure, with $PG14^{RML}$ forming larger, more tightly packed aggregates. We found that the majority of $PG14^{RML}$ aggregates have a sedimentation coefficient of >50S, with ~30% of them having a sedimentation coefficient of >120S. If composed exclusively of PrP, the latter aggregates would contain >200 molecules of the protein. In contrast, 15-20% of $PG14^{Spon}$ PrP was monomeric (3.2S), with the rest sedimenting at 16-20S (corresponding to oligomers containing ~20-30 molecules of PrP). These results suggest that while highly aggregated polymers of PrP are a prerequisite for prion infectivity, small, β-rich oligomers are the species that are toxic to nerve cells (Chiesa et al., 2003). This conclusion is strikingly reminiscent of recent work in Alzheimer's disease, which has demonstrated that small oligomers of the Aβ peptide (ranging in size from 2-100 subunits), rather than large amyloid fibers, are the primary toxic species (Haass and Selkoe, 2007; Walsh and Selkoe, 2007).

In a more recent study (Biasini et al., 2008), we have subjected $PG14^{Spon}$ and $PG14^{RML}$ to a panel of assays commonly used to distinguish infectious PrP (PrP^{Sc}) from cellular PrP (PrP^{C}), including immobilized metal affinity chromatography, precipitation with sodium phosphotungstate, and immunoprecipitation with PrP^{C}- and PrP^{Sc}-specific antibodies. Surprisingly, we found that aggregates of $PG14^{Spon}$ and $PG14^{RML}$ behave identically to each other, and to authentic PrP^{Sc}, in each of these biochemical assays. $PG14^{Spon}$ however, in contrast to $PG14^{RML}$ and PrP^{Sc}, was unable to seed the misfolding of PrP^{C} in an *in vitro* protein misfolding cyclic amplification reaction. Collectively, these results suggest that infectious and non-infectious aggregates of PrP share common structural features accounting for their toxicity, and that self-propagation of PrP involves more subtle molecular differences that remain to be identified.

Gain vs. loss of function as a pathogenic mechanism

How do mutant PrP molecules induce neurodegeneration? For most dominantly inherited neurodegenerative disorders, a gain-of-function mechanism is invoked in which the mutant protein is presumed to acquire a toxic property unrelated to its normal physiological activity (Winklhofer et al., 2008). For example, aggregates of the mutant protein could disrupt important cellular processes and so impair neuronal viability.

Such a mechanism appears to play a role in the pathogenicity of the PG14 PrP, based on our studies of the localization of this mutant in neurons in the brains of transgenic mice. We generated transgenic mice expressing PG14-EGFP, a fluorescent fusion protein that can be directly visualized *in vivo* (Medrano et al., 2008). Tg(PG14-EGFP) mice develop an ataxic neurological illness characterized by astrogliosis, PrP aggregation, and accumulation of a partially protease-resistant form of the mutant PrP. Strikingly, PG14-EGFP forms numerous fluorescent aggregates in the neuropil and white matter of multiple brain regions. These aggregates are particularly prominent along axonal tracts in both brain and peripheral nerve, and similar intracellular deposits are visible along the processes of cultured neurons. Our results suggest that intra-axonal aggregates of mutant PrP could contribute to the pathogenesis of familial prion disease by disrupting axonal transport. A similar mechanism of axonal blockage has recently been proposed to explain the pathogenesis of Alzheimer's disease (Stokin et al., 2005).

To investigate whether a loss-of-function mechanism could also play a role in the neurotoxicity of PG14 PrP, we crossed Tg(PG14) mice with Tg(PrPΔ32-134) mice (Li et al., 2007c). Tg(PrPΔ32-134) mice, which express an N-terminally truncated form of PrP, spontaneously develop a neurodegenerative phenotype that is stoichiometrically reversed by co-expression of wild-type PrP. We found that, at equivalent expression levels, PG14 PrP is significantly less efficient than wild-type PrP in suppressing the development of clinical symptoms and neuropathology in Tg(PrPΔ32-134) mice. Thus, our results suggest that some features of the neurological illness associated with inherited PrP mutations may be attributable to a loss of PrP neuroprotective function.

Since PrP-null mice do not display features of a prion disease (Büeler et al., 1992; Manson et al., 1994), it is unlikely that these disorders are due exclusively to the simple loss of an essential physiological function of PrP^{C}. A loss of function mechanism also appears to be incompatible with the dominant mode of inheritance of familial prion diseases. However, these facts do not rule out a contribution of PrP functional deficiency to prion-related neurodegeneration. For example, PrP^{Sc} or mutant PrP may sequester wild-type PrP^{C} into aggregates that lack functional activity, thereby producing a dominant-negative effect. Moreover a biological activity of PrP^{C} that is dispensable under normal conditions may become essential in the disease state due to cellular or organismal stress.

Taken together, our results also indicate that more than one pathogenic mechanism (both gain- and loss-of-function) may contribute to neurotoxicity in inherited prion diseases.

Topological variants of PrP

There is evidence that alterations in membrane topology can result in neurotoxic forms of PrP. While most PrP molecules are linked to the plasma membrane exclusively via a GPI anchor, three topological variants have been described: ${}^{Cyto}PrP$, ${}^{Ntm}PrP$, and ${}^{Ctm}PrP$ (Chakrabarti et al., 2009). ${}^{Cyto}PrP$, in which the polypeptide chain lies entirely in the cytoplasm, is produced at low levels as a result of inefficient translocation into the ER (Drisaldi et al., 2003). The amount of ${}^{Cyto}PrP$ can be regulated by cellular stress, a mechanism that has been termed "pre-emptive quality control" (Kang et al., 2006). Mice expressing a transgenically encoded form of ${}^{Cyto}PrP$ show a severe neurodegenerative phenotype, indicating that this molecule possesses neurotoxic activity (Ma et al., 2002). It has been claimed that prion infection increases production of ${}^{Cyto}PrP$ (Rane et al., 2008), thereby contributing to neurotoxicity, but this observation is not confirmed by our own data (Stewart and Harris, 2003). At present, the pathogenic role of ${}^{Cyto}PrP$ in prion diseases remains unclear.

Two transmembrane variants of PrP have also been described. ${}^{Ctm}PrP$ and ${}^{Ntm}PrP$ span the lipid bilayer once via a stretch of hydrophobic amino acids in the central region of the protein (residues 111-135), with the N- or C-terminus, respectively, in the cytoplasm. These molecules were first observed in cell-free translation/translocation systems, where it was shown that their amounts could be altered by mutations in the central hydrophobic domain, as well as in an adjacent "stop transfer effector" segment that contains several positively charged amino acids (Hay et al., 1987; Hegde et al., 1998; Yost et al., 1990). Subsequent work from our laboratory (Stewart and Harris, 2003) and others (Kim et al., 2001) demonstrated that non-conservative substitutions in the core of the N-terminal signal sequence also increased the proportion of ${}^{Ctm}PrP$, and that combining these mutations with ones in the central domain resulted in almost exclusive synthesis of the ${}^{Ctm}PrP$. Cell biological analysis of one such compound mutant (L9R/3AV) (Fig. 1) led to the conclusion that ${}^{Ctm}PrP$ contains an uncleaved signal peptide as well as a GPI anchor, and that it is retained intracellularly in either the ER or the Golgi (depending on the cell type) (Stewart et al., 2001; Stewart and Harris, 2005).

Transgenic mice expressing PrP molecules with ^{Ctm}PrP-promoting mutations display a neurodegenerative phenotype, implying that ^{Ctm}PrP has neurotoxic potential *in vivo* (Hegde et al., 1998; Stewart et al., 2005). PrP^{Sc} accumulation has been claimed to cause enhanced generation of ^{Ctm}PrP, and on this basis it has been suggested that ^{Ctm}PrP represents a key neurotoxic intermediate in prion disorders (Hegde et al., 1999). However, this hypothesis is called into question by our observations that scrapie infection does not significantly alter ^{Ctm}PrP levels in cultured cells or brain (Stewart and Harris, 2003), and that pathogenic mutations outside of the central, hydrophobic domain do not alter the membrane topology of PrP (Stewart and Harris, 2001).

Whether or not ^{Ctm}PrP itself plays a role in naturally occurring prion diseases, it is likely to be revealing PrP-related neurotoxic signaling mechanisms that could contribute to the pathological process. Our studies of Tg(L9R/3AV) mice exemplify this point. These animals develop a fatal neurological illness characterized by ataxia and neuronal loss in the cerebellum and hippocampus (Stewart et al., 2005). Importantly, this phenotype is strongly dependent on co-expression of endogenous, wild-type PrP, suggesting a synergistic interaction between transmembrane and exclusively GPI-anchored forms of PrP in transducing a toxic signal.

Neurotoxicity of PrP deletion mutants

One of the most surprising and intriguing observations to have emerged concerning the neurotoxic effects of PrP was first made over 10 years ago by Shmerling et al. (1998) as part of a study to examine which parts of the PrP molecule were necessary for sustaining prion infection. These authors discovered that mice expressing PrP harboring either of two large, N-terminal deletions (Δ32-121 and Δ32-134) (Fig. 1) developed a spontaneous neurodegenerative illness even without inoculation with scrapie prions. This illness was characterized by ataxia and massive degeneration of cerebellar granule neurons. Mice with shorter deletions (Δ32-80, Δ32-93, Δ32-106) were phenotypically normal, implying a critical role for amino acids distal to residue 106. What was even more surprising, the neurodegenerative phenotype of the Tg(Δ32-121) and Tg(Δ32-134) mice was only observed on the *Prn-p$^{0/0}$* background: co-expression of endogenous, wild-type PrP from even a single *Prn-p* allele completely abrogated clinical symptoms and neuropathology. To explain these puzzling observations, the authors proposed a loss-of-function model based on the existence of two hypothetical molecules, one of which (L_{prp}) is a ligand that binds to PrP and the other (π) a receptor that binds the ligand when PrP is absent. It was suggested that binding of L_{prp} to wild-type PrP delivered an essential trophic signal, while binding to N-terminally deleted PrP failed to elicit this signal and also prevented interaction with π.

To more precisely map the region of PrP responsible for the spontaneous neurodegenerative phenotype, we created Tg(ΔCR) mice expressing PrP with a much smaller deletion, comprising residues 105-125 within the Central Region of the molecule (Li et al., 2007b) (Fig. 1). The deleted segment lies within a highly conserved, unstructured region of the protein, and encompasses a cluster of three positively charged amino acids (residues 105, 109, and 110) followed by a stretch of 15 hydrophobic residues (residues 111-135). There were three motivations for focusing on the 105-125 region as a possible toxicity-determining domain in PrP. First, the study by Shmerling et al. (1998) had pinpointed residues distal to position 106 as being critical determinants of neurotoxicity. Second, it was known that a synthetic peptide derived from this region (PrP106-126) was toxic when applied to cultured neurons from *Prn-p$^{+/+}$* but not from *Prn-p$^{0/0}$* mice (Brown et al., 1994a). Third, the 105-125 region is one of the most highly conserved portions of the sequence, being virtually identical in PrP molecules from zebrafish to humans (Rivera-Milla et al., 2006).

Tg(ΔCR) mice display a spontaneous neurodegenerative phenotype that is more severe than any we have ever encountered in other transgenic mice (Li et al., 2007b). This phenotype is observed at modest expression levels of the mutant protein (0.5-1X). On the *Prn-p$^{0/0}$* background, Tg(ΔCR) mice become ill and die within one week of birth. This phenotype is reversed in a dose-dependent fashion by co-expression of wild-type PrP: one *Prn-p* allele (0.5X expression level) delays death until 25 days, and two *Prn-p* alleles (1X expression level) delays death until 48 days. The presence of one Tga20 allele (5X expression level of wild-type PrP) strongly rescues the neurodegenerative phenotype, with mice remaining alive for well over one year.

Tg(ΔCR) mice show two major kinds of neuropathology (Li et al., 2007b). First, there is massive degeneration of cerebellar granule cells (with no effect on Purkinje cells), resulting in severe atrophy of the cerebellum. Degenerating granule neurons display extensive DNA fragmentation, but there is no activation of caspases 3 or 8 (Christensen et al., manuscript in preparation). By electron microscopy, granule cells display a unique, non-apoptotic ultrastructural morphology (Christensen et al., manuscript in preparation). In conjunction with evidence that neurodegeneration in Tg(ΔCR) mice is Bax-independent (Li et al., 2007a), our ultrastructural and biochemical data indicate that PrPΔCR is activating a novel form of neuronal death that is not related to apoptosis, necrosis, or autophagy.

The second major neuropathological feature of Tg(ΔCR) mice is vacuolar degeneration of white matter areas of the brain and spinal cord, particularly in older animals. This abnormality, which was also seen Tg(Δ32-134) mice (Radovanovic et al., 2005), may be due either to axonal or myelin damage.

The biochemical and cell biological properties of PrPΔCR suggest that the neurotoxicity of this molecule results from an alteration of a normal activity of PrP^{C}, rather than from accumulation of misfolded protein aggregates or cellular mislocalization. PrPΔCR is not aggregated or protease-resistant like PrP^{Sc} (Li et al., 2007b). Moreover, the cellular localization of PrPΔCR is identical to that of wild-type PrP (Christensen and Harris, 2009). Like its wild-type counterpart, ΔCR PrP is present in lipid rafts on the plasma membrane, where it is attached by a GPI anchor, and it selectively localizes to the apical surface of polarized MDCK epithelial cells. Moreover, PrPΔCR is distributed in a punctate pattern on the surface of neuronal processes, similar to the wild-type protein.

Recently, Baumann et al. (Baumann et al., 2007) described lines of mice expressing PrP with a deletion of residues 94-134. These animals also display a neurodegenerative phenotype that is suppressible by co-expression of wild-type PrP. It is instructive to compare the features of mice expressing the four different pathogenic deletions thus far published (Δ32-121, Δ32-134, Δ94-134, and Δ105-125) (Fig. 1). Each of these deletions encompasses the 105-125 region, demonstrating the importance of these residues in determining the neurotoxic activity of PrP. Interestingly, the four kinds of mice differ in the severity of the phenotypes they display (as indicated by age at symptom onset), and there is a correlation between the severity of the phenotype and the amount of wild-type PrP required for rescue (Table 1). Tg(ΔCR) mice, which have the strongest phenotype (illness onset at <1 week on the *Prn-p*$^{0/0}$ background), require >5X expression of wild-type PrP for complete rescue, while Tg(PrPΔ34-134) mice (illness onset at 3-5 weeks) are almost completely rescued by 0.5X co-expression of wild-type PrP. This quantitative relationship between neurotoxic potency and the ability to be suppressed by wild-type PrP provides important clues to the underlying molecular mechanisms (see below).

Doppel and shadoo

Studies of two proteins with structural similarities to PrP have provided additional insights into the neurotoxic activities of PrP. Doppel (Dpl) is a PrP paralog that resembles the C-terminal half of PrP, but lacks residues homologous to the flexible N-terminal domain (Mo et al., 2001) (Fig. 1). Dpl thus structurally resembles the deleted forms of PrP that produce neurodegeneration in transgenic mice. The Dpl gene, which is normally expressed primarily in testis, is expressed ectopically in the brain in certain lines of *Prn-p*$^{0/0}$ mice as a result of intergenic splicing events between the adjacent PrP and Dpl genes (Moore et al., 1999). These lines, as well as transgenic lines expressing elevated levels of Dpl in the brain, display a neurodegenerative phenotype that is stoichiometrically rescued by wild-type PrP (Moore et al., 2001; Rossi et al., 2001). Dpl-induced neuronal death involves Purkinje cells as well as granule cells in the cerebellum, and is partially Bax-independent (Dong et al., 2007; Heitz et al., 2007). Taken together, the available data make it very likely that the neurotoxicity of Dpl involves the same molecular pathways as those activated by deleted forms of PrP.

While Dpl is structurally related to the C-terminal half of PrP, the recently described protein, Shadoo (Sho), resembles a GPI-anchored version of the flexible, N-terminal tail of PrP, including the highly conserved central region (reviewed in Watts and Westaway, 2007) (Fig. 1). Sho is expressed in many of the same regions of the brain as PrP^{C}, although the cellular distribution of the two proteins is complementary (Watts et al., 2007). In the cerebellum, for example, Sho is most abundant in Purkinje cell bodies and dendrites, while PrP^{C} is present at highest levels in granule neurons and their axons in the molecular layer. In a neuronal culture assay, Sho had no toxic activity on its own, but was able to suppress the toxicity of co-expressed Dpl and PrPΔ32-121 (Watts et al., 2007). Thus, Sho displays a neuroprotective activity like that of wild-type PrP^{C}. The protective activity of PrP^{C} presumably resides in the N-terminal half of the protein, based on the observation that fusion of this region to Dpl prevents neurodegeneration in transgenic mice (Yoshikawa et al., 2008). Interestingly, Sho protein levels are dramatically down-regulated in scrapie-infected mice, raising the possibility that loss of Sho neuroprotection may play a role in prion diseases (Watts et al., 2007). It has been suggested that Sho is a candidate for π, the hypothetical receptor that has been postulated to deliver an essential neurotrophic signal in the absence of PrP (Watts et al., 2007).

A model for PrP neurotoxicity: subversion of function

Studies of the biological activities of ΔCR and other deleted forms of PrP, along with Dpl and Sho, provide powerful mechanistic insights into the normal function of PrP^{C}, and how it can be subverted to produce neurotoxic effects. The dose-dependent, opposing effects of truncated and wild-type PrP are most easily explained by supposing that both proteins compete for binding to a common molecular target, with binding of truncated PrP eliciting a neurotoxic signal, and binding of wild-type PrP silencing the signal (or restoring a non-toxic, physiological signal) (Fig. 2). The molecular target for PrP is presumably a receptor or other cell-surface complex capable of transducing a signal to the interior of the cell. We hypothesize that binding to the target involves at least two sites on PrP: the CR region (residues 105-125), as well as a more C-terminal region. In the absence of the CR region (as in PrPΔCR), binding of the C-terminal region alone would elicit a neurotoxic signal. Binding at both sites (wild-type PrP) would produce a physiological signal or no signal. We suggest that an identical mechanism is responsible for the neurotoxicity of all four deleted forms of PrP as well as Dpl, with variations in the toxic potency of these molecules being attributable to differences in their affinity for the putative binding target. The cytoprotective effect of Sho could be due to its ability to occupy the CR binding site on the putative transducer molecule.

Although ΔCR and other deleted forms of PrP are artificial molecules, working out how they kill neurons will likely provide crucial insights into neurodegeneration in naturally occurring prion diseases. Like PrPΔCR, PrP^{Sc} (or other toxic forms of PrP that accumulate during prion infection) appears to act by subverting a normal function of PrP^{C} (see above). It is therefore possible that PrP^{Sc}/PrP^{toxic} interacts with the same putative membrane target as PrPΔCR, resulting in generation of a similar neurotoxic signal (Fig. 2). Interestingly, there is evidence that PrP^{Sc} is conformationally altered in the CR region (Peretz *et al*, 1997), perhaps preventing binding interactions involving this site. Thus, PrP^{Sc}/PrP^{toxic}, like PrPΔCR, may be able to interact with its target only via C-terminal binding sites. The same may be true for PG14 PrP, which is aggregated and conformationally altered in the CR region (Biasini et al., 2008) (Fig. 2).

The model outlined here may also explain the toxicity of ^{Ctm}PrP, as well as the synthetic peptide PrP106-126. PrP106-126 has been reported to be toxic to neurons cultured from *Prn-p*$^{+/+}$ but not *Prn-p*$^{0/0}$ mice (Brown et al., 1994a; Fioriti et al., 2005; Forloni et al., 1993). This result suggests that the peptide alters interaction between PrP and the hypothetical transducer by competitively blocking binding within the 105–125 region of PrP. This would then produce a toxic signal equivalent to the one elicited by PrPΔCR which lacks the 105–125 domain. In the absence of PrP^{C}, no signal would be delivered and the peptide would have no effect. In ^{Ctm}PrP, the CR region is partially embedded in the lipid bilayer, and so would be unavailable

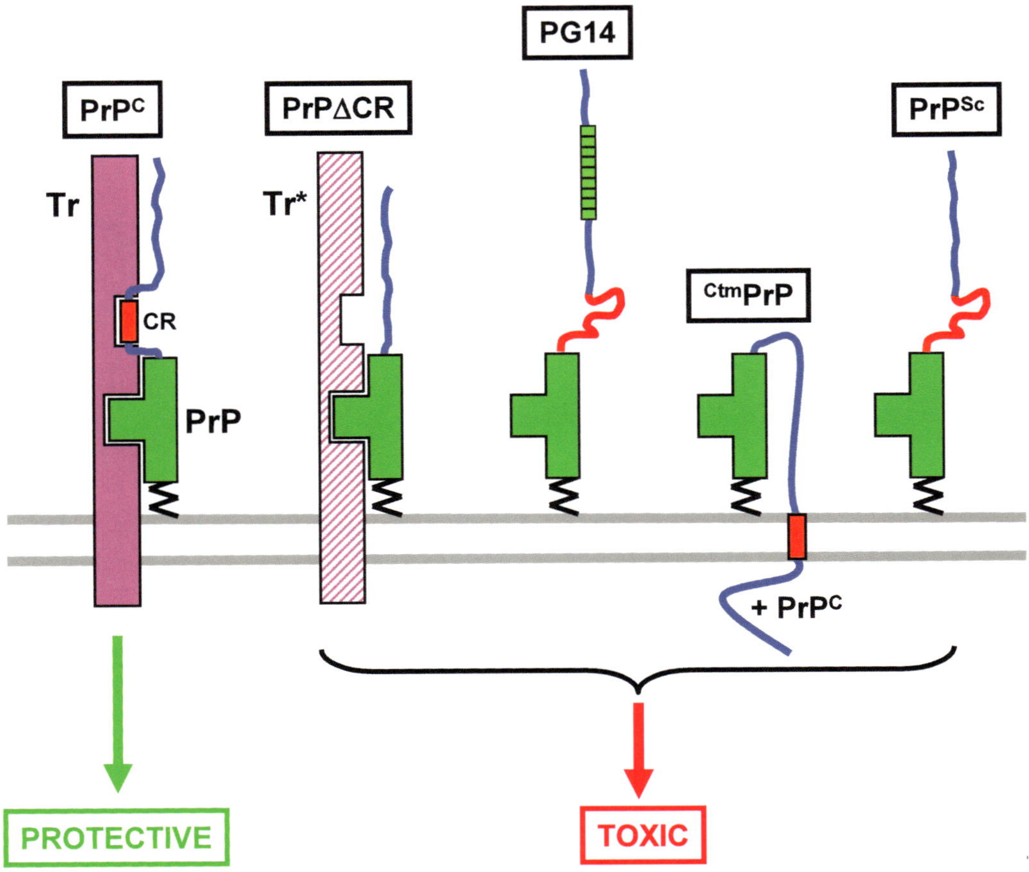

Figure 2. Model for the neuroprotective activity of PrP^{C}, and subversion of this activity by neurotoxic forms of PrP. The structured, C-terminal half of PrP is shown in green and the flexible, N-terminal tail as a blue line. The CR segment of PrP (residues 105–125) is shown as a red rectangle. Tr, hypothetical signal transducing protein that normally generates a neuroprotective signal (solid pink), but which can assume an altered conformation (crosshatched pink) that generates a neurotoxic signal. Two binding sites between PrP and Tr are shown, one involving the C-terminal half of PrP and the other CR segment of PrP. When both binding sites are occupied, Tr elicits a non-essential neuroprotective signal (PrP^{C}). When only the C-terminal site is occupied, as would be the case when the CR segment is absent (PrPΔCR), embedded in the lipid bilayer (^{Ctm}PrP), or conformationally altered (PG14 and PrP^{Sc}), the transducer delivers a neurotoxic signal. The toxicity of ^{Ctm}PrP requires the cooperation of wild-type PrP^{C}.

for binding to the transducer (Fig. 2). In contrast to the situation for PrPΔCR, however, activation of the transducer by ^{Ctm}PrP is presumably facilitated by the presence of wild-type PrP, which potentiates the toxicity of ^{Ctm}PrP.

Conclusions and perspective

Table 1 summarizes the properties of the non-infectious, neurotoxic mutants of PrP discussed in this review, and compares them to the properties of PrP^{C} and infectious PrP^{Sc}. Studies of neurotoxic PrP mutants have provided a number of important insights into the pathogenic species and cellular pathways that may underlie prion-induced neurodegeneration. An important principle exemplified by all of these mutants is the distinction between prion infectivity and neurotoxicity. For some of the mutants associated with familial prion disorders, such as octapeptide insertions, small, β-rich aggregates are likely to represent a key neurotoxic intermediate. There is evidence that these aggregates act via both gain- and loss-of-function mechanisms. In contrast, transmembrane mutants as well as mutants harboring deletions within the central region display biochemical and cell biological properties indistinguishable from those of wild-type PrP. The neurotoxicity of these mutants is likely to depend on alteration or subversion of a neuroprotective function of PrP^{C}.

Further understanding prion neurotoxicity will require progress on several fronts. First, it will be necessary to identify the molecular components of the cellular signaling pathways activated by neurotoxic forms of PrP. This will involve searching for PrP-interacting proteins via biochemical methods, as well as via genetic means using model organisms. The availability of highly potent neurotoxic forms like PrPΔCR should facilitate this search, since these molecules should have an especially high affinity for putative cell surface transducers. Second, it will be important to develop cell culture systems that allow the neurotoxic effects of PrP to be studied *in vitro*. Unfortunately, the mutant PrP molecules discussed here, all of which are neurotoxic in transgenic mice, have only modest effects on cell viability when expressed in cultured cells (our unpublished data) (Ashok and Hegde, 2008; Drisaldi et al., 2004; Rambold et al., 2008; Watts et al., 2007). Presumably, some factor present in the brain milieu required for maximal toxicity of these molecules must be missing in cell culture systems. Finally, and perhaps most importantly, we will need to demonstrate a connection between the neurotoxic mechanisms activated by artificial mutants like PrPΔCR and those operative in "natural" prion diseases of humans and animals. The most direct way to achieve this goal will be to knock down expression of the relevant signaling components *in vivo*, and see if this mitigates the disease phenotype.

Most current approaches to treatment of prion diseases are based on inhibiting accumulation of PrP^{Sc} (Trevitt and Collinge, 2006). Identification of the cellular pathways activated by neurotoxic forms of PrP would allow development of an entirely new class of anti-prion therapeutics based on blocking these pathways. Assuming appropriate cell culture models can be developed, some of the mutant forms of PrP described in this review may provide particularly good substrate for a drug screening approach to identify inhibitors of neurotoxic pathways. Such compounds may have wide applicability, given some of the underlying similarities between prion diseases and other neurodegenerative disorders in terms of pathogenic mechanisms. Indeed, recent evidence that PrP^C is a cell-surface receptor for the Alzheimer's Aβ peptide provides a direct link between neurotoxic mechanisms involved in these two diseases (Laurén et al., 2009).

Acknowledgements

Work in the Harris laboratory is supported by grants from the National Institutes of Health (NS052526 and NS040975) and the Dana Foundation.

References

Aguzzi, A. (2005). Cell biology. Prion toxicity: all sail and no anchor. Science *308*, 1420-1421.

Aguzzi, A., Baumann, F., and Bremer, J. (2008). The prion's elusive reason for being. Annu Rev Neurosci *31*, 439-477.

Ashok, A., and Hegde, R.S. (2008). Retrotranslocation of prion proteins from the endoplasmic reticulum by preventing GPI signal transamidation. Mol Biol Cell *19*, 3463-3476.

Baumann, F., Tolnay, M., Brabeck, C., Pahnke, J., Kloz, U., Niemann, H.H., Heikenwalder, M., Rülicke, T., Bürkle, A., and Aguzzi, A. (2007). Lethal recessive myelin toxicity of prion protein lacking its central domain. EMBO J *26*, 538-547.

Biasini, E., Medrano, A.Z., Thellung, S., Chiesa, R., and Harris, D.A. (2008). Multiple biochemical similarities between infectious and non-infectious aggregates of a prion protein carrying an octapeptide insertion. J Neurochem *104*, 1293-1308.

Bounhar, Y., Zhang, Y., Goodyer, C.G., and LeBlanc, A. (2001). Prion protein protects human neurons against Bax-mediated apoptosis. J Biol Chem *276*, 39145-39149.

Brandner, S., Isenmann, S., Raeber, A., Fischer, M., Sailer, A., Kobayashi, Y., Marino, S., Weissmann, C., and Aguzzi, A. (1996). Normal host prion protein necessary for scrapie-induced neurotoxicity. Nature *379*, 339-343.

Brown, D.R., Herms, J., and Kretzschmar, H.A. (1994a). Mouse cortical cells lacking cellular PrP survive in culture with a neurotoxic PrP fragment. Neuroreport *5*, 2057-2060.

Brown, D.R., Nicholas, R.S., and Canevari, L. (2002). Lack of prion protein expression results in a neuronal phenotype sensitive to stress. J Neurosci Res *67*, 211-224.

Brown, P., Gibbs, C.J., Rodgers-Johnson, P., Asher, D.M., Sulima, M.P., Bacote, A., Goldfarb, L.G., and Gajdusek, D.C. (1994b). Human spongiform encephalopathy: the National Institutes of Health series of 300 cases of experimentally transmitted disease. Ann Neurol *35*, 513-529.

Büeler, H., Fischer, M., Lang, Y., Fluethmann, H., Lipp, H.-P., DeArmond, S.J., Prusiner, S.B., Aguet, M., and Weissmann, C. (1992). Normal development and behavior of mice lacking the neuronal cell-surface PrP protein. Nature *356*, 577-582.

Chakrabarti, O., Ashok, A., and Hegde, R.S. (2009). Prion protein biosynthesis and its emerging role in neurodegeneration. Trends Biochem Sci *doi:10.1016/j.tibs.2009.03.001*.

Chesebro, B., Trifilo, M., Race, R., Meade-White, K., Teng, C., LaCasse, R., Raymond, L., Favara, C., Baron, G., Priola, S., *et al.* (2005). Anchorless prion protein results in infectious amyloid disease without clinical scrapie. Science *308*, 1435-1439.

Chiesa, R., Drisaldi, B., Quaglio, E., Migheli, A., Piccardo, P., Ghetti, B., and Harris, D.A. (2000). Accumulation of protease-resistant prion protein (PrP) and apoptosis of cerebellar granule cells in transgenic mice expressing a PrP insertional mutation. Proc Natl Acad Sci USA *97*, 5574-5579.

Chiesa, R., and Harris, D.A. (2001). Prion diseases: what is the neurotoxic molecule? Neurobiol Dis *8*, 743-763.

Chiesa, R., Piccardo, P., Ghetti, B., and Harris, D.A. (1998). Neurological illness in transgenic mice expressing a prion protein with an insertional mutation. Neuron *21*, 1339-1351.

Chiesa, R., Piccardo, P., Quaglio, E., Drisaldi, B., Si-Hoe, S.L., Takao, M., Ghetti, B., and Harris, D.A. (2003). Molecular distinction between pathogenic and infectious properties of the prion protein. J Virol *77*, 7611-7622.

Christensen, H.M., and Harris, D.A. (2008). Prion protein lacks robust cytoprotective activity in cultured cells. Mol Neurodegener *3*, 11.

Christensen, H.M., and Harris, D.A. (2009). A deleted prion protein that is neurotoxic *in vivo* is localized normally in cultured cells. J Neurochem *108*, 44-56.

Diarra-Mehrpour, M., Arrabal, S., Jalil, A., Pinson, X., Gaudin, C., Pietu, G., Pitaval, A., Ripoche, H., Eloit, M., Dormont, D., *et al.* (2004). Prion protein prevents human breast carcinoma cell line from tumor necrosis factor alpha-induced cell death. Cancer Res *64*, 719-727.

Dong, J., Li, A., Yamaguchi, N., Sakaguchi, S., and Harris, D.A. (2007). Doppel induces degeneration of cerebellar Purkinje cells independently of Bax. Am J Pathol *171*, 599-607.

Drisaldi, B., Coomaraswamy, J., Mastrangelo, P., Strome, B., Yang, J., Watts, J.C., Chishti, M.A., Marvi, M., Windl, O., Ahrens, R., *et al.* (2004). Genetic mapping of activity determinants within cellular prion proteins: N-terminal modules in PrP^C offset pro-apoptotic activity of the Doppel helix B/B' region. J Biol Chem *279*, 55443-55454.

Drisaldi, B., Stewart, R.S., Adles, C., Stewart, L.R., Quaglio, E., Biasini, E., Fioriti, L., Chiesa, R., and Harris, D.A. (2003). Mutant PrP is delayed in its exit from the endoplasmic reticulum, but neither wild-type nor mutant PrP undergoes retrotranslocation prior to proteasomal degradation. J Biol Chem *278*, 21732-21743.

Duchen, L.W., Poulter, M., and Harding, A.E. (1993). Dementia associated with a 216 base pair insertion in the prion protein gene: clinical and neuropathological features. Brain *116*, 555-567.

Fioriti, L., Quaglio, E., Massignan, T., Colombo, L., Stewart, R.S., Salmona, M., Harris, D.A., Forloni, G., and Chiesa, R. (2005). The neurotoxicity of prion protein (PrP) peptide 106-126 is independent of the expression level of PrP and is not mediated by abnormal PrP species. Mol Cell Neurosci *28*, 165-176.

Flechsig, E., Shmerling, D., Hegyi, I., Raeber, A.J., Fischer, M., Cozzio, A., von Mering, C., Aguzzi, A., and Weissmann, C. (2000). Prion protein devoid of the octapeptide repeat region restores susceptibility to scrapie in PrP knockout mice. Neuron *27*, 399-408.

Forloni, G., Angeretti, N., Chiesa, R., Monzani, E., Salmona, M., Bugiani, O., and Tagliavini, F. (1993). Neurotoxicity of a prion protein fragment. Nature *362*, 543-546.

Frigg, R., Wenzel, A., Samardzija, M., Oesch, B., Wariwoda, H., Navarini, A.A., Seeliger, M.W., Tanimoto, N., Reme, C., and Grimm, C. (2006). The prion protein is neuroprotective against retinal degeneration *in vivo*. Exp Eye Res *83*, 1350-1358.

Haass, C., and Selkoe, D.J. (2007). Soluble protein oligomers in neurodegeneration: lessons from the Alzheimer's amyloid beta-peptide. Nat Rev Mol Cell Biol *8*, 101-112.

Harris, D.A. (2003). Trafficking, turnover and membrane topology of PrP. Br Med Bull *66*, 71-85.

Harris, D.A., and True, H.L. (2006). New insights into prion structure and toxicity. Neuron *50*, 353-357.

Hay, B., Barry, R.A., Lieberburg, I., Prusiner, S.B., and Lingappa, V.R. (1987). Biogenesis and transmembrane orientation of the cellular isoform of the scrapie prion protein. Mol Cell Biol *7*, 914-920.

Hegde, R.S., Mastrianni, J.A., Scott, M.R., Defea, K.A., Tremblay, P., Torchia, M., DeArmond, S.J., Prusiner, S.B., and Lingappa, V.R. (1998). A transmembrane form of the prion protein in neurodegenerative disease. Science *279*, 827-834.

Hegde, R.S., Tremblay, P., Groth, D., DeArmond, S.J., Prusiner, S.B., and Lingappa, V.R. (1999). Transmissible and genetic prion diseases share a common pathway of neurodegeneration. Nature *402*, 822-826.

Heitz, S., Lutz, Y., Rodeau, J.L., Zanjani, H., Gautheron, V., Bombarde, G., Richard, F., Fuchs, J.P., Vogel, M.W., Mariani, J., *et al.* (2007). BAX contributes to Doppel-induced apoptosis of prion-protein-deficient Purkinje cells. Dev Neurobiol *67*, 670-686.

Hill, A.F., Joiner, S., Linehan, J., Desbruslais, M., Lantos, P.L., and Collinge, J. (2000). Species-barrier-independent prion replication in apparently resistant species. Proc Natl Acad Sci USA *97*, 10248-10253.

Hoshino, S., Inoue, K., Yokoyama, T., Kobayashi, S., Asakura, T., Teramoto, A., and Itohara, S. (2003). Prions prevent brain damage after experimental brain injury: a preliminary report. Acta Neurochir Suppl *86*, 297-299.

Ivanova, L., Barmada, S., Kummer, T., and Harris, D.A. (2001). Mutant prion proteins are partially retained in the endoplasmic reticulum. J Biol Chem *276*, 42409-42421.

Jeffrey, M., Goodsir, C.M., McGovern, G., Barmada, S., Medrano, A.Z., and Harris, D.A. (2009). Prion protein with an insertional mutation accumulates on axonal and dendritic

plasmalemma, and is associated with distinctive ultrastructural changes. Am J Pathol *In Press*.

Kang, S.W., Rane, N.S., Kim, S.J., Garrison, J.L., Taunton, J., and Hegde, R.S. (2006). Substrate-specific translocational attenuation during ER stress defines a pre-emptive quality control pathway. Cell *127*, 999-1013.

Kim, S.J., Rahbar, R., and Hegde, R.S. (2001). Combinatorial control of prion protein biogenesis by the signal sequence and transmembrane domain. J Biol Chem *276*, 26132-26140.

Krasemann, S., Zerr, I., Weber, T., Poser, S., Kretzschmar, H., Hunsmann, G., and Bodemer, W. (1995). Prion disease associated with a novel nine octapeptide repeat insertion in the PRNP gene. Mol Brain Res *34*, 173-176.

Kuwahara, C., Takeuchi, A.M., Nishimura, T., Haraguchi, K., Kubosaki, A., Matsumoto, Y., Saeki, K., Yokoyama, T., Itohara, S., and Onodera, T. (1999). Prions prevent neuronal cell-line death. Nature *400*, 225-226.

Lasmezas, C.I., Deslys, J.P., Robain, O., Jaegly, A., Beringue, V., Peyrin, J.M., Fournier, J.G., Hauw, J.J., Rossier, J., and Dormont, D. (1997). Transmission of the BSE agent to mice in the absence of detectable abnormal prion protein. Science *275*, 402-405.

Laurén, J., Gimbel, D.A., Nygaard, H.B., Gilbert, J.W., and Strittmatter, S.M. (2009). Cellular prion protein mediates impairment of synaptic plasticity by amyloid-beta oligomers. Nature *457*, 1128-1132.

Lehmann, S., and Harris, D.A. (1996). Mutant and infectious prion proteins display common biochemical properties in cultured cells. J Biol Chem *271*, 1633-1637.

Li, A., Barmada, S.J., Roth, K.A., and Harris, D.A. (2007a). N-terminally deleted forms of the prion protein activate both Bax-dependent and Bax-independent neurotoxic pathways. J Neurosci *27*, 852-859.

Li, A., Christensen, H.M., Stewart, L.R., Roth, K.A., Chiesa, R., and Harris, D.A. (2007b). Neonatal lethality in transgenic mice expressing prion protein with a deletion of residues 105-125. EMBO J *26*, 548-558.

Li, A., and Harris, D.A. (2005). Mammalian prion protein suppresses Bax-induced cell death in yeast. J Biol Chem *280*, 17430-17434.

Li, A., Piccardo, P., Barmada, S.J., Ghetti, B., and Harris, D.A. (2007c). Prion protein with an octapeptide insertion has impaired neuroprotective activity in transgenic mice. EMBO J *26*, 2777-2785.

Lorenz, H., Windl, O., and Kretzschmar, H.A. (2002). Cellular phenotyping of secretory and nuclear prion proteins associated with inherited prion diseases. J Biol Chem *277*, 8508-8516.

Ma, J., Wollmann, R., and Lindquist, S. (2002). Neurotoxicity and neurodegeneration when PrP accumulates in the cytosol. Science *298*, 1781-1785.

Malaga-Trillo, E., Solis, G.P., Schrock, Y., Geiss, C., Luncz, L., Thomanetz, V., and Stuermer, C.A. (2009). Regulation of embryonic cell adhesion by the prion protein. PLoS Biol *7*, e55.

Mallucci, G., Dickinson, A., Linehan, J., Klohn, P.C., Brandner, S., and Collinge, J. (2003). Depleting neuronal PrP in prion infection prevents disease and reverses spongiosis. Science *302*, 871-874.

Mange, A., Milhavet, O., Umlauf, D., Harris, D., and Lehmann, S. (2002). PrP-dependent cell adhesion in N2a neuroblastoma cells. FEBS Lett *514*, 159-162.

Manson, J.C., Clarke, A.R., Hooper, M.L., Aitchison, L., McConnell, I., and Hope, J. (1994). 129/Ola mice carrying a null mutation in PrP that abolishes mRNA production are developmentally normal. Mol Neurobiol *8*, 121-127.

Manson, J.C., Jamieson, E., Baybutt, H., Tuzi, N.L., Barron, R., McConnell, I., Somerville, R., Ironside, J., Will, R., Sy, M.S., *et al.* (1999). A single amino acid alteration (101L) introduced into murine PrP dramatically alters incubation time of transmissible spongiform encephalopathy. EMBO J *18*, 6855-6864.

McLennan, N.F., Brennan, P.M., McNeill, A., Davies, I., Fotheringham, A., Rennison, K.A., Ritchie, D., Brannan, F., Head, M.W., Ironside, J.W., *et al.* (2004). Prion protein accumulation and neuroprotection in hypoxic brain damage. Am J Pathol *165*, 227-235.

Mead, S. (2006). Prion disease genetics. Eur J Hum Genet *14*, 273-281.

Medrano, A.Z., Barmada, S.J., Biasini, E., and Harris, D.A. (2008). GFP-tagged mutant prion protein forms intra-axonal aggregates in transgenic mice. Neurobiol Dis *31*, 20-32.

Mo, H., Moore, R.C., Cohen, F.E., Westaway, D., Prusiner, S.B., Wright, P.E., and Dyson, H.J. (2001). Two different neurodegenerative diseases caused by proteins with similar structures. Proc Natl Acad Sci USA *98*, 2352-2357.

Moore, R.C., Lee, I.Y., Silverman, G.L., Harrison, P.M., Strome, R., Heinrich, C., Karunaratne, A., Pasternak, S.H., Chishti, M.A., Liang, Y., *et al.* (1999). Ataxia in prion protein (PrP)-deficient mice is associated with upregulation of the novel PrP-like protein doppel. J Mol Biol *292*, 797-817.

Moore, R.C., Mastrangelo, P., Bouzamondo, E., Heinrich, C., Legname, G., Prusiner, S.B., Hood, L., Westaway, D., DeArmond, S.J., and Tremblay, P. (2001). Doppel-induced cerebellar degeneration in transgenic mice. Proc Natl Acad Sci USA *98*, 15288-15293.

Mouillet-Richard, S., Ermonval, M., Chebassier, C., Laplanche, J.L., Lehmann, S., Launay, J.M., and Kellermann, O. (2000). Signal transduction through prion protein. Science *289*, 1925-1928.

Owen, F., Poulter, M., Collinge, J., Leach, M., Lofthouse, R., Crow, T.J., and Harding, A.E. (1992). A dementing illness associated with a novel insertion in the prion protein gene. Mol Brain Res *13*, 155-157.

Pauly, P.C., and Harris, D.A. (1998). Copper stimulates endocytosis of the prion protein. J Biol Chem *273*, 33107-33110.

Piccardo, P., Liepnieks, J.J., William, A., Dlouhy, S.R., Farlow, M.R., Young, K., Nochlin, D., Bird, T.D., Nixon, R.R., Ball, M.J., *et al.* (2001). Prion proteins with different conformations accumulate in Gerstmann-Sträussler-Scheinker disease caused by A117V and F198S mutations. Am J Pathol *158*, 2201-2207.

Priola, S.A., and Chesebro, B. (1998). Abnormal properties of prion protein with insertional mutations in different cell types. J Biol Chem *273*, 11980-11985.

Prusiner, S.B. (1998). Prions. Proc Natl Acad Sci USA *95*, 13363-13383.

Prusiner, S.B., ed. (2004). Prion Biology and Diseases, Second edn (Cold Spring Harbor, New York, Cold Spring Harbor Laboratory Press).

Race, R., Raines, A., Raymond, G.J., Caughey, B., and Chesebro, B. (2001). Long-term subclinical carrier state precedes scrapie replication and adaptation in a resistant species: analogies to bovine spongiform encephalopathy and variant Creutzfeldt-Jakob disease in humans. J Virol *75*, 10106-10112.

Radovanovic, I., Braun, N., Giger, O.T., Mertz, K., Miele, G., Prinz, M., Navarro, B., and Aguzzi, A. (2005). Truncated prion protein and Doppel are myelinotoxic in the absence of oligodendrocytic PrP^{C}. J Neurosci *25*, 4879-4888.

Rambold, A.S., Muller, V., Ron, U., Ben-Tal, N., Winklhofer, K.F., and Tatzelt, J. (2008). Stress-protective signalling of prion protein is corrupted by scrapie prions. EMBO J *27*, 1974-1984.

Rane, N.S., Kang, S.W., Chakrabarti, O., Feigenbaum, L., and Hegde, R.S. (2008). Reduced translocation of nascent prion protein during ER stress contributes to neurodegeneration. Dev Cell *15*, 359-370.

Rangel, A., Burgaya, F., Gavin, R., Soriano, E., Aguzzi, A., and Del Rio, J.A. (2007). Enhanced susceptibility of *Prnp*-deficient mice to kainate-induced seizures, neuronal apoptosis, and death: Role of AMPA/kainate receptors. J Neurosci Res *85*, 2741-2755.

Rivera-Milla, E., Oidtmann, B., Panagiotidis, C.H., Baier, M., Sklaviadis, T., Hoffmann, R., Zhou, Y., Solis, G.P., Stuermer, C.A., and Malaga-Trillo, E. (2006). Disparate evolution of prion protein domains and the distinct origin of Doppel- and prion-related loci revealed by fish-to-mammal comparisons. FASEB J *20*, 317-319.

Rossi, D., Cozzio, A., Flechsig, E., Klein, M.A., Rülicke, T., Aguzzi, A., and Weissmann, C. (2001). Onset of ataxia and Purkinje cell loss in PrP null mice inversely correlated with Dpl level in brain. EMBO J *20*, 694-702.

Roucou, X., Gains, M., and LeBlanc, A.C. (2004). Neuroprotective functions of prion protein. J Neurosci Res *75*, 153-161.

Roucou, X., Giannopoulos, P.N., Zhang, Y., Jodoin, J., Goodyer, C.G., and LeBlanc, A. (2005). Cellular prion protein inhibits proapoptotic Bax conformational change in human neurons and in breast carcinoma MCF-7 cells. Cell Death Differ *12*, 783-795.

Roucou, X., and LeBlanc, A.C. (2005). Cellular prion protein neuroprotective function: implications in prion diseases. J Mol Med *83*, 3-11.

Sakurai-Yamashita, Y., Sakaguchi, S., Yoshikawa, D., Okimura, N., Masuda, Y., Katamine, S., and Niwa, M. (2005). Female-specific neuroprotection against transient brain ischemia observed in mice devoid of prion protein is abolished by ectopic expression of prion protein-like protein. Neuroscience *136*, 281-287.

Shmerling, D., Hegyi, I., Fischer, M., Blättler, T., Brandner, S., Götz, J., Rülicke, T., Flechsig, E., Cozzio, A., von Mering, C., *et al.* (1998). Expression of amino-terminally truncated PrP in the mouse leading to ataxia and specific cerebellar lesions. Cell *93*, 203-214.

Spudich, A., Frigg, R., Kilic, E., Kilic, U., Oesch, B., Raeber, A., Bassetti, C.L., and Hermann, D.M. (2005). Aggravation of ischemic brain injury by prion protein deficiency: Role of ERK-1/-2 and STAT-1. Neurobiol Dis *20*, 442-449.

Steele, A.D., Lindquist, S., and Aguzzi, A. (2007). The prion protein knockout mouse: a phenotype under challenge. Prion *1*, 83-93.

Stewart, R.S., Drisaldi, B., and Harris, D.A. (2001). A transmembrane form of the prion protein contains an uncleaved signal peptide and is retained in the endoplasmic reticulum. Mol Biol Cell *12*, 881-889.

Stewart, R.S., and Harris, D.A. (2001). Most pathogenic mutations do not alter the membrane topology of the prion protein. J Biol Chem *276*, 2212-2220.

Stewart, R.S., and Harris, D.A. (2003). Mutational analysis of topological determinants in prion protein (PrP) and measurement of transmembrane and cytosolic PrP during prion infection. J Biol Chem *278*, 45960-45968.

Stewart, R.S., and Harris, D.A. (2005). A transmembrane form of the prion protein is localized in the Golgi apparatus of neurons. J Biol Chem *280*, 15855-15864.

Stewart, R.S., Piccardo, P., Ghetti, B., and Harris, D.A. (2005). Neurodegenerative illness in transgenic mice expressing a transmembrane form of the prion protein. J Neurosci *25*, 3469-3477.

Stokin, G.B., Lillo, C., Falzone, T.L., Brusch, R.G., Rockenstein, E., Mount, S.L., Raman, R., Davies, P., Masliah, E., Williams, D.S., *et al.* (2005). Axonopathy and transport deficits early in the pathogenesis of Alzheimer's disease. Science *307*, 1282-1288.

Tateishi, J., and Kitamoto, T. (1995). Inherited prion diseases and transmission to rodents. Brain Pathol *5*, 53-59.

Tateishi, J., Kitamoto, T., Doh-ura, K., Sakaki, Y., Steinmetz, G., Tranchant, C., Warter, J.M., and Heldt, N. (1990). Immunochemical, molecular genetic, and transmission studies on a case of Gerstmann-Sträussler-Scheinker syndrome. Neurology *40*, 1578-1581.

Tateishi, J., Kitamoto, T., Hoque, M.Z., and Furukawa, H. (1996). Experimental transmission of Creutzfeldt-Jakob disease and related diseases to rodents. Neurology *46*, 532-537.

Trevitt, C.R., and Collinge, J. (2006). A systematic review of prion therapeutics in experimental models. Brain *129*, 2241-2265.

Walsh, D.M., and Selkoe, D.J. (2007). A beta oligomers - a decade of discovery. J Neurochem *101*, 1172-1184.

Watts, J.C., Drisaldi, B., Ng, V., Yang, J., Strome, B., Horne, P., Sy, M.S., Yoong, L., Young, R., Mastrangelo, P., *et al.* (2007). The CNS glycoprotein Shadoo has PrP^{C}-like protective properties and displays reduced levels in prion infections. EMBO J *26*, 4038-4050.

Watts, J.C., and Westaway, D. (2007). The prion protein family: diversity, rivalry, and dysfunction. Biochim Biophys Acta *1772*, 654-672.

Westergard, L., Christensen, H.M., and Harris, D.A. (2007). The cellular prion protein (PrP^{C}): its physiological function and role in disease. Biochim Biophys Acta *1772*, 629-644.

Winklhofer, K.F., Tatzelt, J., and Haass, C. (2008). The two faces of protein misfolding: gain- and loss-of-function in neurodegenerative diseases. EMBO J *27*, 336-349.

Yoshikawa, D., Yamaguchi, N., Ishibashi, D., Yamanaka, H., Okimura, N., Yamaguchi, Y., Mori, T., Miyata, H., Shigematsu, K., Katamine, S., *et al.* (2008). Dominant-negative effects of the N-terminal half of prion protein on neurotoxicity of prion protein-like protein/doppel in mice. J Biol Chem *283*, 24202-24211.

Yost, C.S., Lopez, C.D., Prusiner, S.B., Myers, R.M., and Lingappa, V.R. (1990). Non-hydrophobic extracytoplasmic determinant of stop transfer in the prion protein. Nature *343*, 669-672.

Chapter 2

Prion Protein: Orchestrating Neurotrophic Activities

Vilma R. Martins[1], Flavio H. Beraldo[1], Glaucia N. Hajj[1], Marilene H. Lopes[1], Kil Sun Lee, Marco A. Prado and Rafael Linden

Abstract
PrP^C is highly expressed in both the central and peripheral nervous systems from early stages of development and in adulthood. Its major conformational change and conversion into an abnormal form (PrP^{Sc}) has been associated with the generation of prions, the infectious agent of transmissible spongiform encephalopathies (TSEs). The massive neurodegeneration presented by individuals suffering from these diseases has been associated with the gain of neurotoxic activity of PrP^{Sc}. On the other hand, major neurodegeneration is also observed in transgenic mice expressing PrP^C molecules deleted of specific domains, which points to important functional domains within this molecule, and supports the hypothesis that loss-of PrP^C function may contribute to the pathogenesis of TSEs. Furthermore, a large body of data demonstrates direct or indirect interaction of PrP^C with extracellular matrix proteins, soluble factors, transmembrane proteins, G-protein coupled receptors and ions channels. The ability of PrP^C to drive the assembly of multi-component complexes at the cell surface is likely the basis for its neurotrophic functions. These properties indicate that PrP^C may be relevant for not only the spongiform encephalophaties, but also as an ancillary component of the pathogenesis of other neurodegenerative diseases, and therefore amenable to therapeutic targeting.

The prion protein and the neurotrophic theory

The neurotrophic theory originally referred to the idea that all body organs were dependent on nutritional factors secreted by the nervous system (Levine, 1992). However, and more akin with Cajal´s theories on the neurotrophic activity of the Schwann cell (Azmitia, 2002), the theory gained momentum upon the discovery of the properties of the nerve growth factor - NGF (Levi-Montalcini, 1987). Then, its central tenet became that the survival of neurons during development of the nervous system was determined by the limited availability of trophic molecules delivered by target organs to the nerve terminals (Oppenheim, 1989).

In parallel with the discoveries concerning a wide range of effects of NGF (Levi-Montalcini et al., 1996), work in the 1980s and 1990s greatly expanded the scope of the neurotrophic theory. Adding to the original emphasis on the role of individual target-derived neurotrophins (Thoenen et al., 1993), current neurotrophic theory allows for the operation of multiple neurotrophic systems upon single neurons (Davies, 1996), the production and physiological release of neurotrophic factors by glia (Sariola and Saarma, 2003), the role of afferent-derived molecules (Linden, 1994), trophic effects of both neurotransmitters and neuromodulators (Linden et al., 2005; Martins and Pearson, 2008), as well as cytokines (Krieglstein and Unsicker, 1996; Middleton et al., 2000) and hormones (Datson et al., 2008; Williams, 2008), and encompasses not only the survival but also the differentiation and functional integrity of components of the nervous system, among other complexities (Chao and Bothwell, 2002; Chao, 2003; Kalb, 2005; Levi-Montalcini and Calissano, 2006).

Neurotrophic interactions are based on cell-cell communication through either the release or the presentation at the cell surface, of specific molecules that bind to other molecules present in a target cell. The latter may be either the same cell, providing for autocrine effects, a neighboring cell or even reside at a distance, depending on both the neurotrophic factor and the structure and diffusion properties of the tissue environment. Both cell surface (Huang and Reichardt, 2003; Blochl and Blochl, 2007) and intracellular (Datson et al., 2008), including nuclear (Williams, 2008) receptors, serve as docking elements for the neurotrophic agents, and activate multiple signaling pathways leading to neurotrophic effects. Cell surface receptors constitute the majority of the targets of neurotrophic agents, and they can transfer signals through the plasma membrane either through their own intracellular domains, or via the activation of transmembrane partners (Yano and Chao, 2000; Runeberg-Roos and Saarma, 2007).

Evidence that the prion protein (PrP^C) is involved in neurotrophic activity first appeared from studies in which the expression of PrP^C prevented cell death triggered by serum deprivation of a hippocampal cell line (Kuwahara et al., 1999). It was subsequently shown that engagement of PrP^C with antibodies triggered cell signaling (Mouillet-Richard et al., 2000), and that interaction of PrP^C with a soluble binding partner induced neurotrophic-like effects mediated by the cAMP-protein kinase A and Mitogen activated protein kinase (MAPK) pathways (Chiarini et al., 2002; Zanata et al., 2002; Lopes et al., 2005). Evidence that PrP^C mediates neurotrophic interactions is now both abundant and compelling (e.g. (Nishimura et al., 2004; McLennan et al., 2004; Spudich et al., 2005; Coulpier et al., 2006; Rangel et al., 2007; Nazor et al., 2007; Weise et al., 2008; Rambold et al., 2008).

The hypothesis has been put forward that the prion protein functions as a dynamic cell surface platform for the

[1]These authors contributed equally to this work.

The Prion Protein 2010 Jörg Tatzelt (Ed.) Savanna Press, UK
ISBN: 978-0-9543335-2-2. savannapress.com

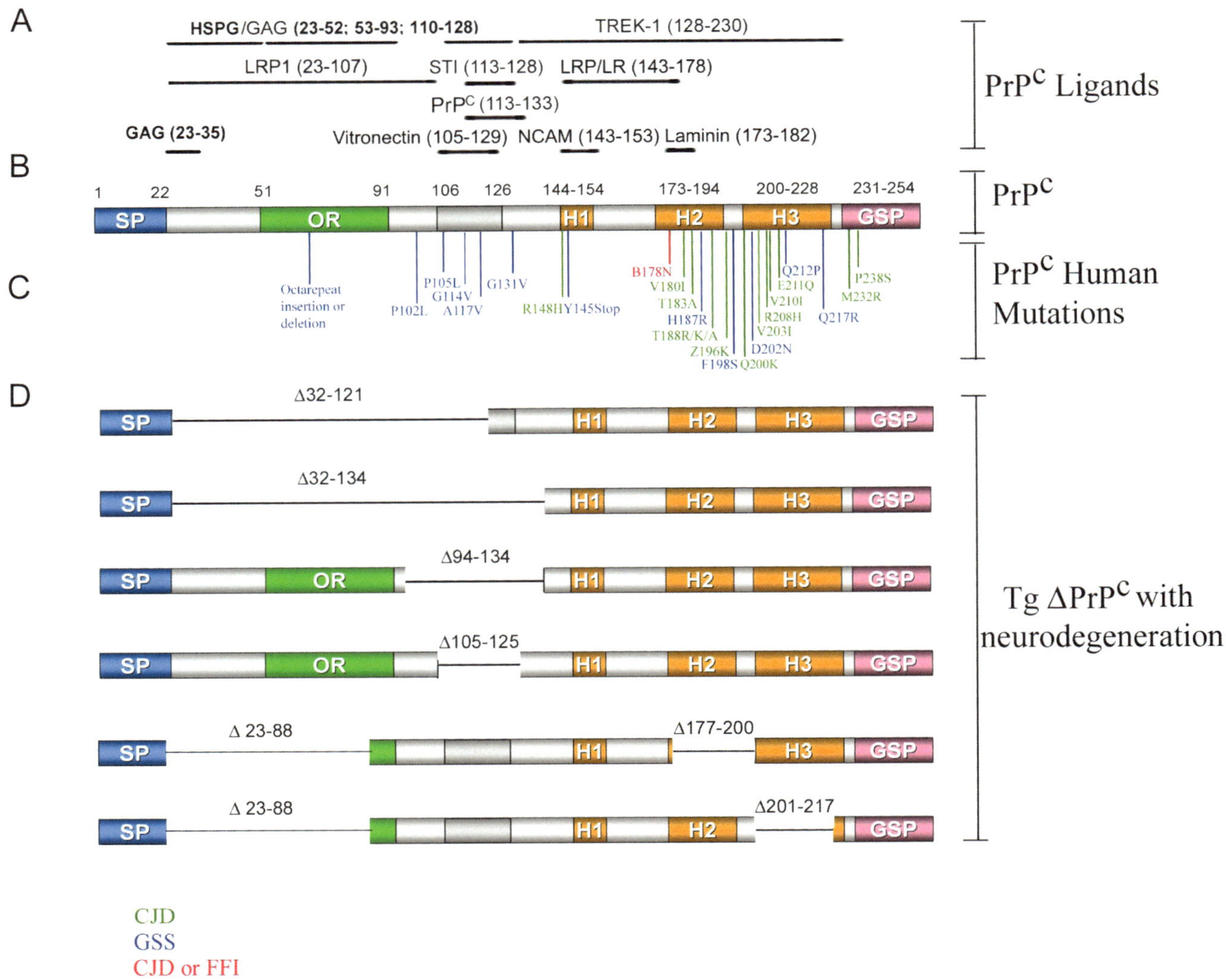

Figure 1. A schematic representation of PrP^C with ligand binding sites, human mutations leading to prion diseases, and transgenic mouse models presenting neurodegeneration.
A) Mouse PrP^C binding sites for ligands with neurotrophic activity. Numbers in parentheses indicate residue position. B) Schematic drawing of mouse PrP^C. During the post-translational modification, the N-terminal signal peptide (SP) and GPI sequence (GPS) are cleaved and a GPI anchor is attached to the C-terminal. The flexible N-terminal portion contains octapeptide repeats (OR) and a hydrophobic domain (dark grey). Orange boxes (H1, H2 and H3) represent regions of α-helical secondary structure. C) Mutations associated with inherited human prion diseases are indicated, CJD (green), GSS (Blue) and GSS/FFI (red). D) Deletions of PrP^C expression which cause neurodegeneration in transgenic mice, Δ numbers indicate the deleted domains.

assembly of signaling modules, based on which selective interactions with many ligands and transmembrane signaling pathways translate into wide-range consequences upon both physiology and behavior (Linden et al., 2008). Thus, PrP^C may scaffold various sets of extracellular and transmembrane molecules, providing for distinct signaling events, depending on the cell type, developmental stage, level of expression of both PrP^C and its partners, the availability of ligands in the immediate environment of the target cells, and endocytic trafficking (Linden et al., 2008). The current review focuses upon the role of PrP^C as a mediator of neurotrophic interactions, in the context of the hypothesis of loss-of-function components of prion diseases.

Following the cleavage of a signal peptide, the mammalian prion protein is produced mostly in the form of an N-glycosylated, GPI-anchored protein of 208-209 aa (Riek et al., 1997; Lopez et al., 2000; Zahn et al., 2000; Hornemann et al., 2004; Calzolai et al., 2005; Lysek et al., 2005; Parkin et al., 2007), containing an N-terminal flexible, random coil sequence that spans approximately half of its aminoacid residues, plus a C-terminal globular domain that constitutes the other half. The globular domain of human PrP^C is arranged in 3 α-helices corresponding to aas 144-154, 173-194 and 200-228, interspersed with an antiparallel β-pleated sheet formed by β-strands at residues 128-131 and 161-164. A single disulfide bond is formed by cysteine residues 179 and 214. The N-terminal residues

23-124 constitute the flexible tail, while residues 229-230 form a short flexible C-terminal domain (Zahn et al., 2000). Major structural features of PrP^C are remarkably preserved among both mammalian and non-mammalian species (Lopez et al., 2000; Hornemann et al., 2004; Calzolai et al., 2005; Lysek et al., 2005).

Full length PrP^C is found in non-, mono- or di-glycosylated forms, corresponding to the variable occupancy of residues Asn181 and Asn197 in human PrP^C - Asn180 and Asn196 in mice (Haraguchi et al., 1989). A rather large variety of N-glycans were found attached to both full-length and truncated PrP^C (Rudd et al., 1999), which may be differentially distributed in various areas of the central nervous system (CNS) (DeArmond et al., 1999; Beringue et al., 2003). Although neither the attached N-glycans nor most of the GPI anchor appear to significantly affect major structural features of the prion protein (Hornemann et al., 2004), both may interfere upon the binding of physiological ligands, which, in turn, may be relevant for its neurotrophic properties (Linden et al., 2008).

PrP^C deletions in mice and mutations in humans: a tip for neurotrophic functions?

Since the early 1990s, knockout constructions for PrP gene (*$Prnp^{0/0}$* or *$Prnp^{-/-}$*) have emerged as an attempt to elucidate the PrP^C functions in vivo (Bueler et al., 1992; Manson et al., 1994). However, the first study in PrP^C-null mice did not reveal an obvious phenotype implying that PrP^C is either not essential for normal development, or could be compensated by other molecules (Bueler et al., 1992; Manson et al., 1994). However, when distinct features of these animals were investigated in further studies, subtle changes have been noticed in certain biological processes such as neurotransmission and synaptic plasticity (Collinge et al., 1994; Maglio et al., 2004; Prestori et al., 2008), hippocampal spatial memory (Criado et al., 2005) and aversive hippocampal memory in aged animals (Coitinho et al., 2003), circadian rhythms (Tobler et al., 1996; Tobler et al., 1997) and immune responses (de Almeida et al., 2005; Bainbridge and Walker, 2005). Remarkably, PrP^C-null mice also show higher sensitivity to various stress conditions which cause increased neuronal death (Walz et al., 1999; McLennan et al., 2004; Weise et al., 2004; Shyu et al., 2005; Weise et al., 2006; Rangel et al., 2007). Therefore, although compensatory mechanisms may balance the absence of PrP^C, they are not sufficient and the threshold for neuronal death is lower when PrP^C is deleted.

Large and diverse phenotypic abnormalities were observed when PrP^C deletion mutants were re-expressed in *$Prnp^{0/0}$* mice. PrP^C mutants deleted either aas 32-121, (PrPΔ32-121) or 32-134 (PrPΔ32-134), led to severe ataxia and apoptosis in the cerebellum (Shmerling et al., 1998). Severe demyelination and axon loss in both the spinal cord and cerebellar white matter were also observed in PrPΔ32-134 mice (Radovanovic et al., 2005) (Figure 1D). The leucodystrophy, but not the cerebellar granule cell degeneration, was rescued in these mice by oligodendrocyte-specific expression of PrP^C (Radovanovic et al., 2005). On the other hand, neuron-specific expression of PrP^C partially rescued cerebellar granule cell degeneration but not demyelination (Radovanovic et al., 2005). This indicates that these diseases are distinct, and that endogenous expression of PrP^C in both neurons and glia was required for complete reversion of the degenerative phenotype.

Transgenic mice expressing truncated PrPΔ94-134, but not those expressing a smaller deletion PrPΔ114-121, also showed extensive central and peripheral myelin degeneration and early ataxia, which progressed to spastic paraparesis and full paraplegia (Baumann et al., 2007). This lethal phenotype was not associated with the presence of pathological aggregates, altered PrP^C glycosylation, subcellular mislocalization or inappropriate membrane topology (Baumann et al., 2007). Interestingly, the most prominent phenotype was observed in animals expressing an even smaller truncation, PrPΔ105-125. These mice developed an extremely severe illness within two weeks of birth, with decreased body size and weight, immobility, impaired righting reflexes, myoclonus and tremor, and died within one month. Histopathology demonstrated cerebellar atrophy, severe loss of cerebellar granule cells, gliosis and astrocytic hypertrophy (Li et al., 2007). It is important to point that the degenerative phenotype observed in mice expressing PrPΔ32-121, PrPΔ32-134, PrPΔ94-134 or PrPΔ105-125 was rescued by introducing the wild-type *Prnp* gene (Shmerling et al., 1998; Baumann et al., 2007; Li et al., 2007).

These deleted regions likely contain functional PrP^C domains relevant for trophic interactions. They lay within the N-terminal and hidrophobic regions, which are the less structured domains and therefore more accessible regions. In fact, this is the interface for PrP^C dimerization (Rambold et al., 2008) and the region where most of the binding sites for PrP^C ligands such as Heparan Sulphate Proteoglicans (HSPG), vitronectin, low-density lipoprotein receptor related protein 1 (LRP1) and Stress Inducible Protein 1 (STI1) were identified (Figure 1A).

Transgenic mice expressing truncated molecules at the C-terminal of PrP^C, also presented an altered phenotype. The expression of truncated PrP^C molecules, PrPΔ177-200 or PrPΔ201-217, associated with another deletion at aas 23 to 88, which is by itself innocuous, caused a neuronal storage disease associated with neurodegeneration and signals of cerebellar disorder (Muramoto et al., 1997). Both deletions caused PrP^C accumulation within cytoplasmatic inclusion bodies in enlarged neurons. Relative smaller number of axons in the white matter tracks were observed in PrPΔ177-200 while PrPΔ201-217 animals presented extensive nerve cells loss in the CA1 region of the hippocampus (Muramoto et al., 1997) (Figure 1D).

The PrP^C ligands at the C-terminal domain of PrP^C characterized until recently are laminin, 37 kDa Laminin Receptor Precursor/ 67 kDa Laminin Receptor (37LRP/67LR), and the K^+ channel, TREK1 (Figure 1A). All these ligands present neurotrophic properties, and particularly the association of laminin with PrP^C allows for neuritogenesis and maintenance of growth cones (see below).

In turn, a large spectrum of clinical symptoms and pathological features are found in patients presenting genetic prion diseases. A series of at least 55 pathogenic mutations are distributed along the PrP^C sequence, and respond for 3 distinct illnesses: Creutzfeldt - Jakob disease (CJD), Gerstmann-Sträussler-Scheinker (GSS) Syndrome, and Fatal Familial Insomnia (FFI) (Figure 1C).

Two octapeptide deletions and 29 insertions were described in the octarepeat region, with the former being associated with a CJD phenotype and later often linked with GSS (Kong Q et al., 2003). Seven point mutations are found in the N-terminal domain and Helix 1 (H1) of PrP^C and 6 of them cause GSS (Figure 1C). On the other hand, 11 out of 17 mutations described in the C-terminal domain of PrP^C cause CJD and most of them are clustered in helices 2 and 3 (H2 and H3) (reviewed by Kong et al., 2003; Rodriguez et al., 2005; Ye et al., 2008). Thus, there is a preferential cluster of GSS phenotypes associated with the N-terminal and H1 regions and a preferential CJD phenotype linked to the C-terminal domain of PrP^C.

The human PrP^C domain comprehending aas 95 to 135 (equivalent to mouse 94 to 134) presents 5 point mutations (P102L, P105L, G114V, A117V and G131V) all of each associated with GSS. This is exactly the PrP^C domain the deletion of which causes neurodegeneration in mice, as discussed above. Remarkably, although brain homogenates obtained from GSS patient with P102L mutation induced spongiform encephalopathy in recipient animals (Masters et al., 1981), mice overexpressing the transgene P101L (corresponding to the human 102) showed spontaneous neurodegeneration without any detectable protease-resistant PrP^{Sc} (Telling et al., 1996). In addition, the mutation A117V which appears to induce the formation of a transmembrane form (Ctm) of PrP^C, produced neurodegenerative changes in mice but does not cause either PrP^{Sc} deposition or infection in rodents (Hegde et al., 1998). Conversely, the expression of the CJD-linked mutations T183A (DeArmond et al., 1997) or E199K (human E200K) (Telling et al., 1996) in PrPnull mice did not cause any pathological signals.

In cell cultures, the PrP^C hydrophobic domain 113-123 was required for the stress-protective activity of PrP^C (Rambold et al., 2008). In addition, some PrP^C mutants associated to genetic prion diseases and presenting codon 129M are unable (D178, V181I, E196K, V210I, M232R and P238S) or only partially effective (R208H, E211Q) to protect against Bax-mediated cell death (Jodoin et al., 2007). Interestingly, the presence of 129V renders mutated proteins into less active forms (A117V, V203I and T188A) or inactive (E200K, R208H, E211Q) for this phenptype (Jodoin et al., 2007).

The truncation of specific PrP^C domains in transgenic mice or the presence of *PRNP* mutations associated with human genetic prion diseases may alter the recruitment of specific proteins to a PrP^C-based multi-component complex (Martins et al., 2002; Linden et al., 2008) in particular those already known to interact with specific PrP^C domains and promote neurotrophic functions. In addition, the composition of these complexes may vary among neurons and glial cells which may explain, at least in part, the diverse degenerative phenotypes observed in transgenic mice and in patients expressing mutated PrP^C.

The PrP^C ligands mediating neurotrophic activity

Several dozen proteins have been described to bind PrP^C in various cellular compartments, but in most cases there is still no known associated biological function for these interactions (Linden et al., 2008). PrP^C is normally localized at the cell surface and traffics along the endocytic pathway (Prado et al., 2004), consistent with the idea that its association with other proteins in these compartments may be relevant for its neurotrophic activity.

Extracellular matrix (ECM) components, as well as their receptors at the cell surface, are known to create a milieu necessary for survival and differentiation in the developing nervous system (Pires Neto et al., 1999; De and Georges-Labouesse, 2000) and in the remodeling of adult brain in both normal (Grimpe and Silver, 2002) and pathological conditions (Rauch, 2004; Theodosis et al., 2004). PrP^C is able to associate with ECM proteins such as laminin (Graner et al., 2000a; Graner et al., 2000b) and vitronectin, but not to fibronectin or collagen (Hajj et al., 2007). PrP^C also binds to ECM glycosaminoglycans such as heparin and heparan sulfate (Warner et al., 2002) and to ECM receptors such as the 37 kDa Laminin Receptor Precursor/ 67 kDa Laminin Receptor (37LRP/67LR) (Rieger et al., 1997).

Laminins are large (400–900 kDa) extracellular heterotrimeric glycoproteins composed of various combinations of α, β, and γ chains. To date, five α, four β, and three γ chains have been identified, which may combine into sixteen known forms of laminins in mammals. The individual chains present diverse biological active domains and interact with a large number of receptors, such as at least eight different integrins, 37LRP/67LR, dystroglycan, 67 kDa elastin-laminin receptor, galactoside-binding lectin, galactosyltransferase, heparan sulfate proteoglycans and immunoglobulin-related basal cell adhesion molecule (Tzu and Marinkovich, 2008). These varied interactions are crucial to promote the diverse biological functions attributed to laminin in distinct cell types, particularly in the nervous system (Colognato and Yurchenco, 2000). For example, laminin is one of the most important ECM proteins responsible for axonal growth (Luckenbill-Edds, 1997). In addition, defects in both the peripheral and central nervous system frequently accompany mutations in laminin isoforms. Also, the loss of interneuronal laminin in the hippocampus due to proteolysis induced by excitotoxicity, leads to neuronal death (Colognato et al., 2005).

PrP^C binds to a laminin domain at the γ-1 chain (Graner et al., 2000a; Graner et al., 2000b), a region previously demonstrated to promote neuritogenesis (Liesi et al., 1989). The laminin binding site resides within amino acids 173-182 of the PrP^C molecule (Coitinho et al., 2006). Using primary hippocampal neuron cultures from either wild-type or PrP^C-null mice, it was shown that the interaction of PrP^C with the laminin γ-1 chain induces neuritogenesis (Graner et al., 2000a). Inactivation of PrP^C by chromophore-assisted laser inactivation (CALI) in PC-12 cells also impaired neurite extension induced by laminin. Furthermore, retraction of growth cones was observed when micro-CALI was used to inactivate PrP^C in these structures (Graner et al., 2000b).

The importance of the interaction of PrP^C and laminin was also demonstrated *in vivo*. Antibodies against either the entire PrP^C and laminin molecules, or against peptides representing the binding sites either in PrP^C or in laminin γ-1 chain, blocked memory consolidation when infused in the rat hippocampus. In addition, the γ-1 chain peptide competed with antibodies against the PrP^C 170-183 peptide, and reversed the latter's inhibitory activity upon memory consolidation (Coitinho et al., 2006). Thus, PrP^C-aminin interaction in the hippocampus is associated with neuritogenesis, with an impact upon neuronal plasticity and memory consolidation.

PrP^C also interacts with 37LRP/67LR, one of the first ligands of PrP^C identified by using a two-hybrid system

(Rieger et al., 1997). The 37LRP/67 LR is a membrane associated protein, first discovered by laminin–sepharose affinity chromatography. At first, 37LRP/67LR interaction with laminin was considered to be simply adhesive. However, it is now clear that activation of the 37LRP/67LR induces dynamic events (increased filopodia, directional motility and modulation of gene expression), as a consequence of hitherto poorly understood signal transduction. In addition to direct interactions with laminin, it has been proposed that the 67LR facilitates interactions between laminin and integrins (Nelson et al., 2008).

The interaction between PrP^C and the 37LRP/67LR was confirmed in mammalian cells, where it was shown that 37LRP/67LR co-localizes with PrP^C at the cell surface. Recombinant or purified PrP^C bind 37kDa LRP/ 67kDa LR at the cell surface, where the latter mediates the internalization of exogenous added PrP^C (Gauczynski et al., 2001). It should be noted, however, that this exogenous added PrP^C lacks the localization in rafts and is of course not GPI-anchored at the membrane. The binding residues for 37kDa LRP/ 67kDa LR have been identified between amino acids 144-179 in PrP^C (Hundt et al., 2001), which partially overlaps the binding site for laminin itself (Coitinho et al., 2006). Thus, it is likely that interaction of PrP^C either with the 37LRP/ 67LR or laminin may be mutually exclusive at this PrP^C domain. Although this competitive interaction has yet to be tested, such an overlap, together with the coincidence of binding sites in 37LRP/ 67LR for both laminin and PrP^C (Rieger et al, 1999), should have important consequences for the effects of PrP^C upon neurite growth and differentiation.

Binding of 37LRP/67LR to an additional binding site at PrP^C amino acids 53 to 93 is mediated by the presence of heparan sulfate proteoglycans at 37LRP/67LR. Interestingly, the 37LRP/67LR domain which contains heparan sulfate proteoglycans maps to between amino acids 101–160 or 180–295 (Hundt et al., 2001), the latter of which straddles a laminin binding region (205-229) (Nelson et al., 2008). Indeed, two out of three laminin binding sites at the 37LRP/67LR may compete by the latter's interaction with PrP^C.

Remarkably, barring steric hindrance constrains, laminin may be able to simultaneous bind the 37LRP/67LR (using its third available binding site located within the cysteine-rich EGF-like repeat in the β1 subunit) (Nelson et al., 2008), PrP^C (at the C-terminal portion of the γ1 chain), as well as other receptors, such as integrins (most of which bind to the α chain of laminin) (Tzu et al., 2008). These multiple interactions would lead to the assembly of a molecular cluster of cell membrane receptors, which may promote neurotrophic signaling by both integrin-mediated signal transduction, as well as through signals initiated by activation of PrP^C by laminin, followed by its possible internalization mediated by the 37LRP/67LR (Figure 2A). This is but one of the multicomponent signaling modules that may be assembled through molecular interactions mediated by the prion protein (Martins et al., 2002; Linden et al., 2008).

Interaction of PrP^C with glycosaminoglycans, particularly heparan sulfate, has been extensively documented (Priola and Caughey, 1994). In addition, *in vitro* binding studies demonstrated that PrP^C directly interacts with heparan sulfate through domains 23–52, 53–93, and 110–128 (Warner et al., 2002). Although no functional consequence of this interaction has been addressed so far, it may have important consequences especially for the binding and internalization of the 37LRP/ 67LR.

The PrP^C role as an ECM receptor/co-receptor is not limited to laminin. PrP^C also binds another ECM protein, vitronectin (Vn), through amino acids 106-116 (Hajj et al., 2007). This PrP^C-Vn interaction is important for the growth of axons in dorsal root ganglia during embryogenesis (Hajj et al., 2007) (Figure 2A). Interestingly, vitronectin has been described to have important roles in motor neuron differentiation in association with the morphogen Sonic Hedgehog (Pons and Marti, 2000; Pons et al., 2001). Also during development of the cerebellum, the migration of granular neurons is regulated by an ECM transition. When cells are in the external granule cell layer, the most predominant ECM molecule is laminin, which stimulates granule cells to proliferate. When cells start to migrate to the internal granule cell layer, they move towards an ECM region rich *in vitro*nectin. Cells that reach vitronectin stop proliferating and differentiate, forming the layers of the fully differentiated cerebellum (Pons et al., 2001). Since PrP^C binds both laminin and vitronectin it may be involved in the development of the cerebellum, by regulating the precise moment of each association and thus influencing the fate of the cells.

Remarkably, knocking down PrP^C in cerebellar neural circuits and in particular on granule cells, compromises the function of these cells causing behavioral alterations such as low performance in rotarod and runway tests. It was demonstrated that cerebellar granule cells from wild-type mice stop their division by the second post-natal week, as expected, while PrP^C-null granule cells divide until post-natal week three. With older age, these differences disappear indicating that the absence of PrP^C can be compensated during the development (Prestori et al., 2008).

Perhaps the most extensively studied PrP^C-binding protein, so far, is Stress Inducible Protein 1 (STI1). Interaction of PrP^C with a 66kDa protein was discovered in 1997, and the latter was proposed as a PrP^C receptor, which might be responsible for mediating PrP^C internalization and inhibiting the toxicity of the PrP^C peptide 106-126 (Martins et al., 1997). It was subsequently found that the interaction between the 66kDa protein and PrP^C triggered protection against apoptosis in an *in vitro* retina model (Chiarini et al., 2002). Notwithstanding the lack of evidence for its functioning as a PrP^C receptor (Zanata et al., 2002), the role of STI1 as an important inducer of PrP^C-mediated neurotrophic effects was further supported by the demonstration that their binding promoted both protection against neuronal death as well as neuritogenesis, through the PKA and ERK1/2 signaling pathways, respectively (Lopes et al., 2005) (Figure 2B).

Recent work in astrocytes has shown that STI1 is a secreted PrP^C ligand. STI1 is released by cultured astrocytes in large amounts by still unknown mechanisms, and interacts with neuronal PrP^C, promoting protection against cell death (Lima et al., 2007). Effects of STI1 *in vivo* have also been demonstrated. Experiments with rats in an inhibitory avoidance task demonstrated that antibodies against STI1 and PrP^C, when injected directly into the hippocampus, impair memory formation and consolidation. Remarkably, the STI1 peptide that represents the PrP^C binding site improves memory formation and consolidation (Coitinho et al., 2007). It is, however, still unknown whether the effects upon memory are related to neurotrophic effects of the interaction of STI1 with PrP^C.

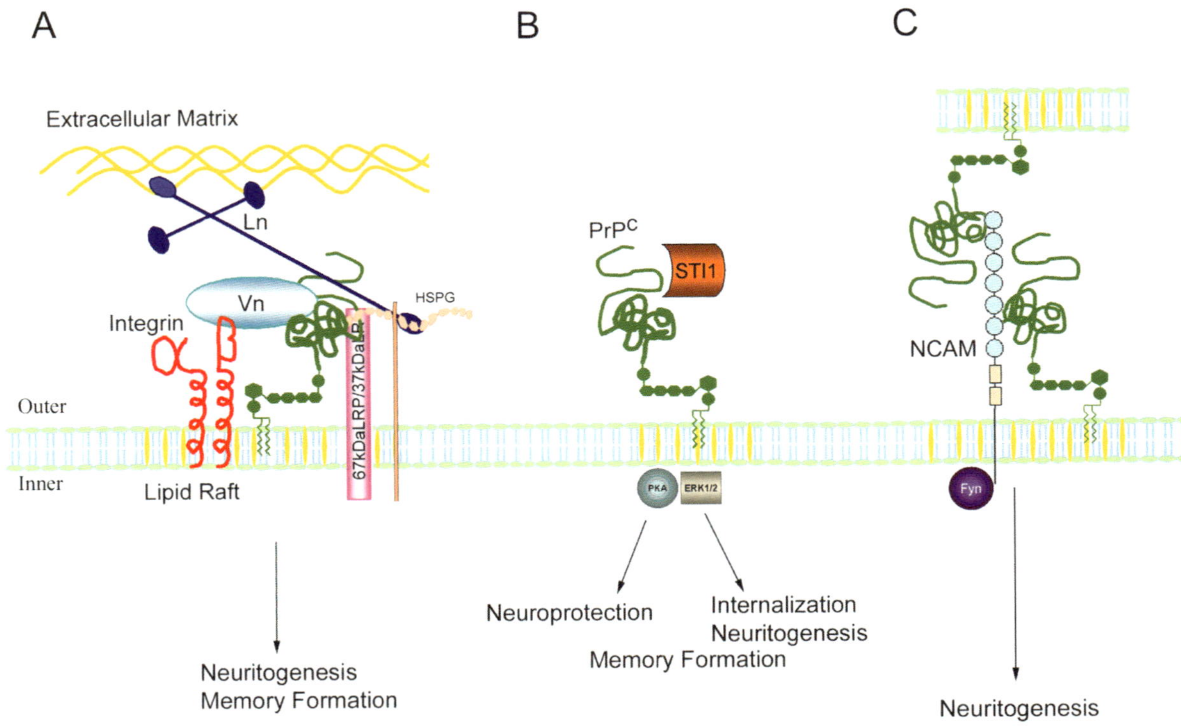

Figure 2. The PrP^{C} ligands mediating neurotrophic activity.
A) PrP^{C} directly interacts with the ECM proteins: Laminin (Ln) and Vitronectin (Vn), and with the laminin receptor (37LRP/67LR), possibly forming a large complex involved in neuritogenesis and memory formation. Interaction of the laminin receptor with PrP^{C} is also mediated by Heparan Sulfate Proteoglycans (HSPG). Integrins are ECM receptors, and may also take part in this complex. B) PrP^{C} interacts with the secreted Stress Inducible Protein 1 (STI1), which mediates neuroprotection and neuritogenesis through independent signaling pathways (PKA and ERK1/2). C) NCAM is a transmembrane protein that interacts with PrP^{C} either in *cis* or in *trans*, and promotes neuritogenesis through the activation of Fyn kinase.

PrP^{C} modulates expression and/or activity of transmembrane receptors associated with neurotrophic activities. Since so many ligands have been ascribed to PrP^{C}, the fundamental question remaining about PrP^{C} is how the signal is transduced through the membrane. PrP^{C} is a GPI-linked protein known to be preferentially localized in membrane domains enriched in sphingolipids and cholesterol (lipid rafts) (Fivaz et al., 2002). These raft domains contain several molecules which recruit specialized proteins for intracellular signaling (Gorodinsky and Harris, 1995) and it is possible that PrP^{C} can modulate the activity of these proteins by either direct or indirect interaction.

In fact, several transmembrane proteins with neurotrophic activities were shown to be modulated by PrP^{C}, leading to the activation or inhibition of intracellular signal pathways. These proteins include the neural cell adhesion molecule (NCAM) (Santuccione et al., 2005), integrins (Hajj et al., 2007), G protein coupled receptors (Kristensson et al., 1993; Wong et al., 1996; Mouillet-Richard et al., 2000; Brini et al., 2005) and membrane ion channels (Whatley et al., 1995; Colling et al., 1996; Herms et al., 2001; Azzalin et al., 2006; Khosravani et al., 2008).

NCAM plays crucial roles in brain development, synaptic plasticity and regeneration and is considered an inducer of complex intracellular signaling cascades in response to extracellular cues triggered by either homophilic or heterophilic binding (Ditlevsen et al., 2008). PrP^{C}-NCAM interaction is direct (Schmitt-Ulms et al., 2001) and occurs in either *cis* or *trans*, i.e. both molecules in the same cell or in distinct cells, respectively. This interaction takes place in raft microdomains, where NCAM uses its PrP^{C}-guided enrichment in lipid rafts to activate p59fyn kinase and inducing neuritogenesis (Santuccione et al., 2005) (Figure 2C). These data are especially interesting, since it has been demonstrated that Fyn kinase pathway is the most relevant for NCAM-induced neurite outgrowth and synaptic functions (Ditlevsen et al., 2008). On the other hand, NCAM-induced signaling also engages other signaling molecules, such as focal adhesion kinase, growth-associated protein-43, the mitogen-activated protein kinase (MAPK), intracellular Ca^{2+}, and protein kinases A, C, and G (reviewed by (Ditlevsen et al., 2008). In fact, it has been recently demonstrated that PrP^{C} accumulates is focal adhesions, modulationg cell-substrate interactions both in cells from mammals and *Drosophila*. When PrP^{C} is highly expressed it modulates cell spreading and filopodia formation while in low levels of PrP^{C} lamellipodia is more abundant (Schrock et al., 2009).

The major role of PrP^{C} in cell adhesion and cell-cell communication has also been elegantly demonstrated in zebrafish which presents two PrP^{C} gene orthologues (PrP1 and PrP2). Deletion of PrP1 caused impairment of embryonic cell adhesion and arrested gastrulation (Malaga-Trillo et al., 2009) while PrP2 removal affects latter stages of neuronal development possibly affecting proliferation and

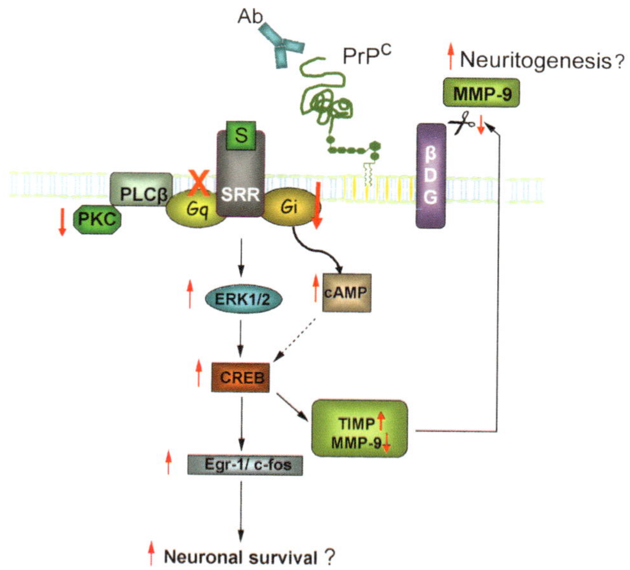

Figure 3. Engagement of PrP^C with specific antibodies modulates signal transduction by altering the activity of G protein-coupled serotonergic receptor. The serotonergic receptor (SRR) in the cell membrane is associated with a calcium stimulatory G protein (Gq), to the Phospholipase Cβ (PLCβ and with a cAMP inhibitory G protein (Gi). This complex responds to serotonin (S) by activating PKC, modulating the levels of cAMP/ERK1/2 (Extracellular Signal-Regulated Kinases 1/2) and CREB (cAMP response element binding protein) pathways. PrP^C binding to specific antibodies (Ab) impaired PLCβ/PKC signaling pathway and partially blocked the reduction of cAMP, up modulates ERK, CREB and Egr-1/c-Fos. These factors may contribute to neuronal survival. On the other hand, up modulation of CREB increases the transcription of TIMP-1 and decreases MMP-9 levels. MMP-9 is a secreted metalloproteinase that cleaves β-dystroglycan (BDG) reducing cell-matrix adhesion. Upon PrP^C ligation, MMP9 levels are diminished and β-dystroglycan interaction with the extracellular matrix is stabilized, what may increase neuritogenesis.

differentiation of developing neurons (Malaga-Trillo et al., 2009 and unpublished results). The alterations related to PrP1 are caused by deficient cell movements modulated by E-cadherin–based adhesion and signaling which are conserved across vertebrates.

As described above PrP^C is able to interact with vitronectin. The vitronectin peptide which represents the PrP^C binding site promotes axonal growth in wild-type but not in PrP^C-null neurons. Vitronectin, on the other hand, supports axonal growth in both wild-type and PrP^C-null neurons, suggesting that other vitronectin ligands can compensate for the absence of PrP^C. In fact, the conduction of functional assays showed that neurons derived from two different PrP^C-null mouse strains presented higher activity of the integrin αvβ3 (Hajj et al., 2007), a classical vitronectin transmembrane receptor. Although, a direct interaction of αvβ3 with PrP^C has never been demonstrated, β3 integrin immunoprecipitates with PrP^C (Hajj G and Martins VR, unpublished results) and β1 integrin is engaged by antibody crosslinking-mediated neuritogenesis (Pantera et al., 2009), indicating that they can associate in the same protein complex (Figure 2A).

Integrins are a large group of heterodimeric surface membrane receptors widely expressed in the nervous system and have been shown to have diverse roles in axon extension, neural development and, in addition, they are key regulators of synaptic plasticity involved in memory and learning in the adult nervous system (reviewed by (Clegg et al., 2003). The negative control of integrin activity by PrP^C may represent an important regulation of neurotrophic functions.

In the neuroectodermal progenitor 1C11 cell line, that can be differentiated into either serotonergic ($1C11^{5\text{-}HT}$) or noradrenergic ($1C11^{NE}$) phenotype, the engagement of PrP^C with antibodies had a high impact on signal transduction (Mouillet-Richard et al., 2000). This effect was restricted to cell processes and most likely on varicosities of the neurites (Mouillet-Richard et al., 2000). PrP^C engagement with G-protein coupled serotonergic receptors (GPCR) 5-HT, 5HT2A, 5HT1B/D and 5-HT2B simultaneously with their stimulation by agonists, altered the intensities and/or dynamics of G protein activation (Mouillet-Richard et al., 2005). In particular, PrP^C binding with antibodies impaired phospholipase Cβ (PLCβ) activation by 5-HT2A receptors, partially blocked the reduction of cAMP promoted by agonist binding to 5-HT1B/D and potentiated the Phospholipase A2 response through 5-HT2A (Mouillet-Richard et al., 2005). In addition, the same procedure triggered transduction cascades controlling the cellular redox state and the ERK1/2 kinases (Mouillet-Richard et al., 2007), leading to CREB activation followed by transcription of Egr-1 and c-fos. These factors have been previouly associated with the control of neuronal survival (Zhang et al., 2002; Thiel and Cibelli, 2002). On the other hand, the transcription of the metalloproteinase TIMP-1 was enhanced by CREB while the transcription of MMP-9 was reduced, decreasing the processing of β-dystroglycan. Since MMP-9 process laminin, which is also a ligand for β-dystroglycan, it has been suggested that the PrP^C-mediated control of MMP-9 stabilizes the interaction between PrP^C and its cell/ECM partners, and may potentiate synaptic efficacy (Pradines et al., 2008) (Figure 3).

The transfection of the PrP^C gene into CHO cells affects intracellular calcium dynamics following activation of purinergic receptors (P2Y). In transfected cells treated with ATP, Ca^{2+} mobilization from endoplasmic reticulum and Ca^{2+} uptake by mitochondria were lower than in non-transfected cells (Brini et al., 2005). These alterations are known to reduce cellular sensitivity to Ca^{2+}-mediated apoptosis and therefore, are cytoprotective (Giacomello et al., 2007; Marchi et al., 2008). In addition, the expression of PrP^C in CHO cells also increased the the Ca^{2+} influx through the plasma membrane after ATP treatment (Brini et al., 2005). It was speculated that the latter effect was mediated by PrP^C activation of store operated calcium channels (SOCCs) (Brini et al., 2005). These channels are functionally coupled with the activation of adenyl cyclase (Fagan et al., 2000) which has been associated with PrP^C-mediated neuronal survival (Chiarini et al., 2002; Lopes et al., 2005) (Figure 4). Whether Ca^{2+} hadling of the main intracellular stores and entry from extracellular space depend on the direct interaction of PrP^C with P2Y receptors or SOCCs is still unknown.

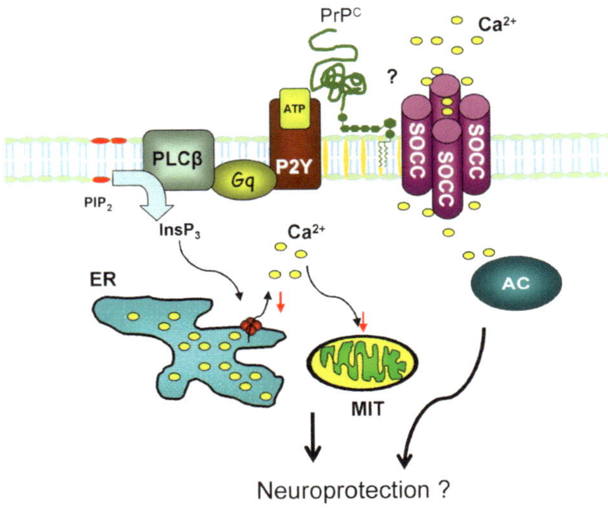

Figure 4. Expression of PrP^C modulates intracellular Ca^{2+} homeostasis and signaling transduction through G protein-coupled purinergic receptors. The purinergic receptors (P2Y) are transmembrane proteins coupled to calcium stimulatory G protein (Gq) and to phospholipase Cβ (PLCβ). P2Y receptors are activated by ATP, promoting cleavage of phosphatidylinositol-bisphosphate (PIP2) into Inositol-3-Phosphate (InsP3) by the PLCβ. InsP3 promotes a calcium release from the endoplasmic reticulum (ER), and released calcium can be sequestered by the mitochondria, leading to cell death. The expression of PrP^C leads to a reduction of Ca^{2+} mobilization from the ER and Ca^{2+} uptake by the mitochondria what can contribute to neuronal protection. In contrast, there is an increasing of calcium influx through the plasma membrane, probably by the activation of Store Operated Calcium Channels (SOCCs) and increase of the activity of adenylate cyclase. The latter has been associated with PrP^C-mediated neuroprotection.

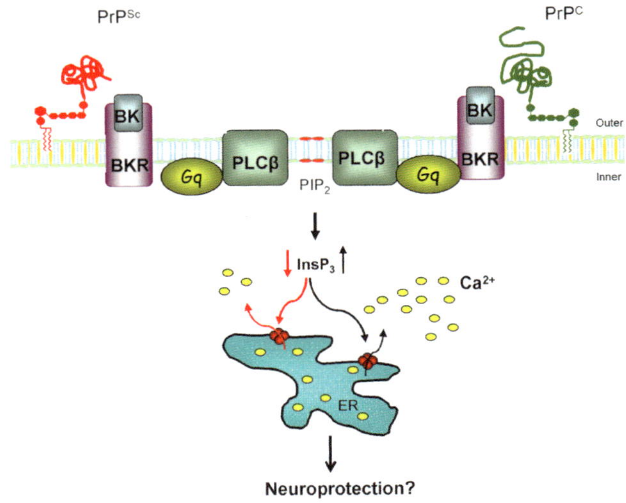

Figure 5. Prion infection impairs mobilization of intracellular Ca^{2+} by bradykinin. The bradykinin receptor (BKR) is a cell membrane protein coupled to calcium stimulatory G protein (Gq) and to Phospholypase Cβ (PLCβ). When bradykinin (BK) binds to its receptor, it promotes the PLCβ-catalysed cleavage of phosphatidylinositolbisphosphate (PIP2) into Inositol-3-Phosphate (InsP3). InsP3 promotes calcium release from endoplasmic reticulum (ER), which has been attributed a role against ischemia. In cells infected with PrP^{Sc}, there is an increase in BKR expression. Paradoxically, there is a reduction of 90% in InsP3 production, resulting in a decrease of Ca^{2+} mobilization from the endoplasmic reticulum (red arrows). These effects are attributed to both reduced affinity of BK for its receptor in this condition, and to a modification of lipid composition in the plasma membrane, with diminished availability of PIP2.

Prion infection reduced the effect of bradykinin on the increase of intracellular Ca^{2+} (Kristensson et al., 1993). This alteration occurs by a reduced affinity of BK for its receptor and also by a modification of lipid composition in the plasma membrane with the diminished availability of phosphatidylinositol-bisphosphate (PIP2). Bradykinin is a 9-amino acid peptide with a wide range of biological actions mediated through B1 and B2 subtypes of G-protein-coupled receptors (reviewed by (Calixto et al., 2004). Activation of these receptors induces a transient increase in the cytosolic concentration of calcium caused by both its mobilization from intracellular stores and influx from extracellular environment (Kristensson et al., 1993 ; Martins et al., 2005). This has been associated with neuroprotection after ischemia (Ping et al., 2005; Yan-Feng et al., 2008; Danielisova et al., 2008). Whether PrP^C modulates the activity of bradykinin receptors, and the suppression of bradykinin activity in prion infected neurons is due to PrP^C-loss-of-function deserves further investigation (Figure 5).

PrP^C also interacts with ionic channels activated by conformational modification promoted by either voltage and/or binding to a specific ligand. Once activated, these proteins promote changes in the concentrations of Na^+, K^+, Ca^{2+} or Cl^-, thus modulating a variety of electrophysiologically-dependent cellular responses. An interaction between PrP^C and a two-pore potassium channel protein, TREK (TWIK-1-related K^+ channel), has been demonstrated using a bacterial two-hybrid approach (Azzalin et al., 2006). TREK-1 (Fink et al., 1996) is a protein that forms a mechanically-gated channel activated by temperature, membrane stretch and internal acidosis, and involved in neuroprotection via activation of PKA (Patel and Honore, 2001; Franks and Honore, 2004). Immunoprecipitation assays confirmed the association between both proteins, and that the binding site for TREK localizes at the carboxy-terminal of PrP^C (aas 128-230). Confocal analysis demonstrated that fluorescent transfected PrP^C and TREK co-localized and, more importantly, that endogenous proteins exhibited co-localization in cerebellar Purkinje cells (Azzalin et al., 2006) (Figure 6A).

The involvement of PrP^C with L-type voltage-gated calcium channels (VGCC) has been first claimed some years ago (Whatley et al., 1995). Reduction of calcium influx, through VGCC, was observed in both cerebellar granule cells and hippocampal neurons slices of PrP^C-null

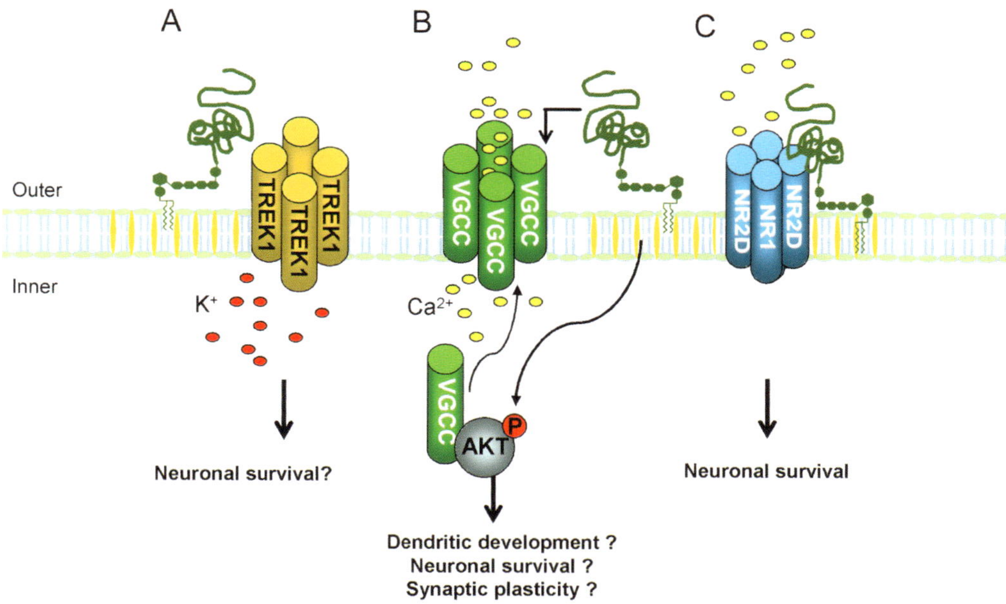

Figure 6. PrP^C interacts with or modulates the activity of ionic channels.
A) PrP^C interacts with a two-pore potassium channel protein (TREK-1), which forms a mechanically-gated K^+ channel, and reportedly promotes neuroprotection via PKA. B) Voltage gated calcium channels (VGCCs) promote an increase in a cytoplasmic calcium. Cells from PrP^C-null mice present a reduction in calcium influx by VGCCs, probably caused by the impairment in AKT activity and phosphorylation of the VGCC subunits which is an essential step for their insertion in the membrane. C) PrP^C interacts with the NR2D subunit of NMDA (N-methyl-D-aspartic acid) receptors, which are ionotropic glutamate receptors permeable to Ca^{2+}. When PrP^C is absent, the NMDA channel is more sensitive to NMDA, which promotes an increase of calcium influx leding to neuronal cell death.

mice when compared to wild-type neurons (Herms et al., 2000; Fuhrmann et al., 2006). In fact, it is known that VGCC are modulated by PI3K/AKT (Viard et al., 2004), whose activation is fundamental to increase VGCC insertion or prolong their expression at the plasma membrane. There is no evidence of a direct interaction of PrP^C with VGCC. However, it is known that activity of PI3K/AKT is diminished PrP^C-null mice (Weise et al., 2006), which may contribute either to a lower insertion or to abbreviate the surface expression of these receptors. One of the most relevant roles of VGCCs is to regulate mechanisms of calcium ions delivery into specific intracellular compartments for a variety of calcium-dependent processes; including dendritic development, neuronal survival, and synaptic plasticity (reviewed by (Moreno, 1999). PrP^C could then modulate neurotrophic functions indirectly through its positive effect on the activation on VGCC (Figure 6B).

Several studies demonstrated that hippocampal neurons from PrP^C-null mice presented increased excitability when compared to wild-type animals (Collinge et al., 1994; Colling et al., 1996; Mallucci et al., 2002). *In vivo* studies also showed that, when compared to wild-type mice, PrP^C-null animals show a lower seizure threshold and higher mortality after treatment with convulsant drugs such as the ionotropic glutamate analog, kainic acid (Walz et al., 1999). Interestingly, PrP^C-null mice presented an elevated susceptibility to neuronal damage induced by kainic acid which is inhibited by the N-methyl-D-aspartate (NMDA) receptor blocker MK-801 (Rangel et al., 2007). NMDA receptors are formed by a tetrameric channel (two NR1 and two NR2 subunits), which are activated by glutamate and are permeable to both sodium and calcium ions (Meldrum, 2000). The expression of NMDA subtypes can be regulated by PrP^C (Maglio et al., 2004). Remarkably, PrP^C associates with NMDA receptors containing NR2D and downregulates their function, preventing depolarization and calcium influx, thus playing a neuroprotective role (Khosravani et al., 2008) (Figure 6C).

PrP^C shedding and secretion: PrP^C as a neurotrophic factor

Despite the preferential PrP^C localization at the cell surface, as a glycosylphosphatidylinositol (GPI)-anchored protein, this molecule presents alternative topologies that may have inherent physiological functions. Soluble forms of PrP^C have been identified not only in the medium of cultured cells (Borchelt et al., 1993; Harris et al., 1993; Parizek et al., 2001; Mattei et al., 2009) but also in human cerebrospinal fluid (Tagliavini et al., 1992), human and murine serum (Parizek et al., 2001; Mattei et al., 2009), and released by human platelets (Perini et al., 1996). At least four mechanisms have been associated to "soluble" forms of PrP^C: 1) release to the extracellular space after removal of its GPI-anchor by post-translational modifications possibly caused by escape of glycolipidation (Borchelt et al., 1993; Harris et al., 1993) (Figure 7A); 2) shedding from the cell surface via phospholipase C cleavage of the GPI anchor (Parkin et al., 2004) (Figure 7A); 3) secretion by exosomes (Fevrier et al., 2004) (Figure 7B) and 4) endoproteolytic cleavage by ADAMs (A Disintegrin And Metalloproteases) (Vincent et al., 2001; Cisse et al., 2005) (Figure 7C).

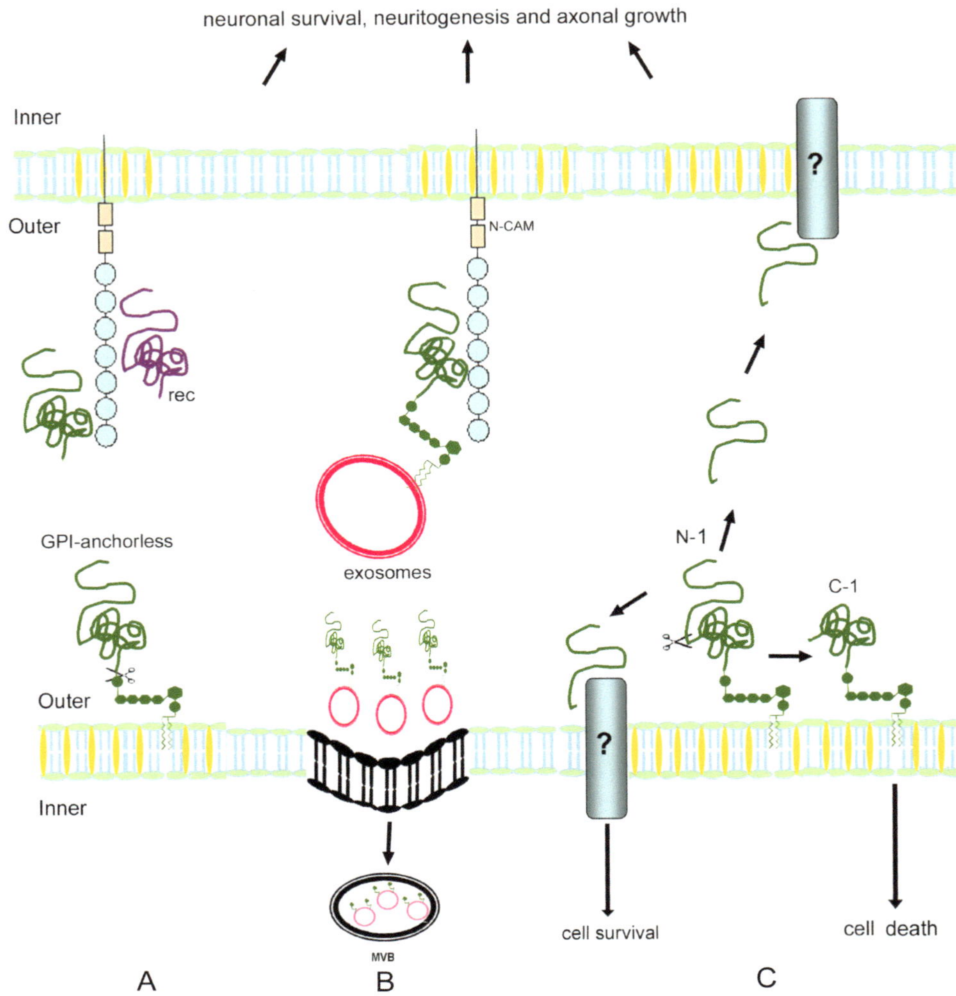

Figure 7. PrP^C can reach the extracellular space either in exosomes or in soluble forms, and mediate neurotrophic activities. PrP^C can be released from cells through distinct mechanisms. A) The PrP^C GPI-anchor can be removed by posttranslational modification or cleaved by phospholipase C. The anchorless PrP^C found in the extracellular space, as well as exogenous recombinant PrP^C, can bind NCAM and modulate neurotrophic functions. B) PrP^C is associated with exosomes, which are released by cells upon fusion of multivesicular bodies (MVB) with the plasma membrane. PrP^C in exosomos can interact with transmembrane proteins, i.e. NCAM, to transduce neuroprotective effects. C) Metaloproteases cleave PrP^C at residues 110/111 generating a soluble N-terminal fragment (N-1) and a C-terminal fragment (C-1) tethered to the plasma membrane. The soluble N-1 fragment induces neuronal survival probably by interacting in either *cis* or *trans* with an unidentified cell surface receptor. The C-1 fragment tethered to the plasma membrane has pronounced pro-apoptotic activity.

Soluble PrP^C can act as a ligand triggering neurotrophic functions, and these functions were mimicked in cell cultures by recombinant protein. When mouse primary cerebellar granule neurons were grown in the presence of recombinant PrP^C, signal transduction components such as PKA, Src-related tyrosin kinases, phosphatidylinositol3-kinaseAkt, and MAPK/Erk kinases were activated, promoting neurite outgrowth and neural survival (Chen et al., 2003). In addition, *in vitro* experiments demonstrated that recombinant PrP^C induces both neuronal survival in mouse cortical neurons (Lima et al., 2007), and rapid neuronal polarization and development of synapses in embryonic rat hippocampal neurons (Kanaani et al., 2005).

Although secreted forms of PrP^C were identified in exosomes and microvesicles from neuronal cells, (Fevrier et al., 2004; Mattei et al., 2009), platelets (Robertson et al., 2006) and ovine cerebral spinal fluid (Vella et al., 2008) their biological functions are unknown. Remarkably, PrP^C secreted from astrocytes has been associated with neuronal survival (Lima et al., 2007), and vesicular structures containing PrP^C were identified by biochemical methods and ultrastructural criteria following immunohistochemistry of astrocyte conditioned media (C. Arantes, G. Hajj, M. H. Lopes, I. Porto-Carreiro, R. Linden and V.R. Martins, unpublished observations) (Figure 7B).

GPI-anchored PrP^C is not a partner for soluble PrP^C, since similar effects were observed when the latter was used to treat either wild-type or PrP^C-null neurons (Chen et al., 2003; Kanaani et al., 2005; Santuccione et al., 2005; Lima et al., 2007). Conversely, NCAM-null neurons were unable to respond to soluble PrP^C (Santuccione et al., 2005), thus indicating that NCAM is required for the neurotrophic functions mediated by physiologically secreted PrP^C molecules or recombinant PrP^C (Figures 7A and B).

PrP^C is endo-proteolytically cleaved at the N-terminal side, at the 110/111 residues, thus producing a 17-kDa C-terminal fragment C1 tethered to the plasma membrane (Harris et al., 1993; Chen et al., 1995; Laffont-Proust et al., 2005). The 9-kDa soluble N-terminal counterpart referred as N-1 is constitutively released in the extracellular medium depending on protein kinase C (Vincent et al., 2000). Three disintegrins contribute to this processing, either directly such as ADAM-10 and -17 (Vincent et al., 2001) or indirectly such as ADAM-9 (Cisse et al., 2005). The activation of muscarinic receptor increases the physiological processing of PrP^C through upregulation of the phosphorylation and activity of ADAM-17 (Alfa et al., 2007).

The C-terminal fragment that remains attached to the cell membrane (C-1) displays pro-apoptotic functions and regulates p53 mRNA transcription and activity (Sunyach et al., 2007) (Figure 7C). Conversely, a recombinant N-1 fragment protected cells from staurosporine-induced apoptosis, by reducing caspase-3 activity, p53 promoter trans-activation and activity as well as p53 mRNA levels (Sunyach in preparation, cited in (Vincent et al., 2008) (Figure 7C). These data indicate that opposite effects of C-1 and N-1 domains may maintain an appropriate balance between cell survival and death.

The mechanisms associated to the pro- and anti-apoptotic effects of the C-1 and N-1 PrP^C fragments remain unknown, and deserve further investigation. These effects may be due to the inability of the membrane-bound C-1 to bind and/or to modulate certain PrP^C partners such as GAG, vitronectin, ligands involved in the neurotrophic role of PrP^C, or LRP1, which binds PrP^C between amino acids 23-107 and is involved with internalization of PrP^C (Parkyn et al., 2008). However, the anti-apoptotic role of PrP^C has been ascribed to the cAMP/PKA pathway (Lopes et al., 2005), and neither the activation of this pathway nor its survival-promoting role appear to be dependent on endocytosis (Caetano et al., 2008). On the other hand, the protective effects of N-1 fragment can occur after its interaction in either *cis* or *trans* with unidentified cell surface proteins (Figure 7C).

Therefore, both GPI-anchored and soluble PrP^C, present overlapping neurotrophic properties. This is consistent with the hypothesis that PrP^C functions as a cell surface scaffolding protein, since the relative preservation of the structure of PrP^C irrespective of its anchoring into the plasma membrane would allow similar molecular interactions of either the membrane-bound or the released forms (Linden et al., 2008). Nonetheless, small structural changes of PrP^C upon interaction with lipids (Morillas et al., 1999; Eberl et al., 2004; Hicks et al., 2006) may affect interaction of PrP^C with its ligands. It is possible that neurotrophic effects of the GPI-anchored PrP^C may be complemented by specific regulation of PrP^C secretion, shedding and cleavage.

PrP^C localization in specific membrane microdomains and PrP^C cellular traffic modulate neurotrophic functions

Lipid rafts form an ordered phase in membranes rich in cholesterol and sphingolipids and are likely to be the sites where GPI-anchored proteins are preferentially partitioned. It has long been recognized that PrP^C, as other GPI-anchored proteins (Fivaz et al., 2002), is present in lipid rafts (Naslavsky et al., 1997). Early work indicated that PrP^{Sc} was present in caveole-like domains in brain tissue (Vey et al., 1996), however pioneer studies using chicken PrP^C showed that the protein is sequestered from the cell surface in clathrin-coated pits (Shyng et al., 1994). This work received some initial criticism, due to the perceived default pathway of internalization of GPI-anchored proteins by caveolae and other non-clathrin mediated pathways (Vey et al., 1996). However, recent work has shown that PrP^C in neurons and neuronal cells is indeed predominantly sequestered by a clathrin-mediated pathway (Sunyach et al., 2003). Both dynamin K44A, a dominant-negative inhibitor of dynamin activity (Magalhaes et al., 2002), and AP-180C, the C-terminal fragment of the endocytic protein AP180, when overexpressed in cells block the internalization of PrP^C (Taylor and Hooper, 2007). Both these treatments have been suggested to block selectively clathrin-mediated endocytosis, albeit dynamin K44A also appears to interfere with caveolae-dependent endocytosis (Prado et al., 2004). Co-localization experiments in neurons shows that mammalian PrP^C is found in clathrin-coated vesicles that also internalize transferrin (Sunyach et al., 2003). In addition, a motif in the N-terminus of cellular prion (^{23}KKRPKP^{28}) appears to be critical for clathrin-mediated endocytosis (Sunyach et al., 2003). These results agree with earlier assessment that found GFP-PrP^C in Rab5-positive endocytic vesicles (Magalhaes et al., 2002).

These observations raise the possibility that PrP^C uses alternative mechanisms other than its GPI-anchor for internalization. A minimum GPI-anchored protein GFP-GPI, utilizes non-clathrin mediated endocytosis as a pathway for internalization (Nichols et al., 2001). Hence, it appears that the GPI-anchor of PrP^C does not contribute majorly for its sequestration from the cell surface. Alternative mechanisms of internalization for GPI-anchored proteins have been suggested. For example, the urokinase plasminogen activator receptor (uPAR, a GPI-anchored protein) piggybacks onto the low-density lipoprotein receptor-related protein (LRP1, a transmembrane scavenger protein that can interact with intracellular adaptor proteins) to be internalized (Nykjaer et al., 1992; Conese et al., 1995; Czekay et al., 2001). Manipulations that alter LRP1 expression or its interaction with ligands (using the chaperone RAP for example) decrease the internalization of PrP^C, suggesting that PrP^C may follow a similar internalization pathway by piggybacking with LRP1 (Taylor et al., 2007; Parkyn et al., 2008). Since PrP^C is found in clathrin-coated vesicles (Sunyach et al., 2003) it might be possible that this is the patway used to internalize PrP^C-LRP1 (Figure 8A). Moreover, LRP1 appears to have additional roles in PrP^C trafficking, as this scavenger receptor appears to interact with PrP^C during its biosynthesis, perhaps acting like a chaperone for PrP^C (Parkyn et al., 2008).

The 37LRP/67LR also promotes PrP^C internalization, even though this has been observed only for the recombinant protein and no internalization pathway has been described (Gauczynski et al., 2001) (Figure 8B).

Recent work from our laboratories suggested that at least for one PrP^C ligand, STI1, their binding triggers endocytosis of the prion protein (Caetano et al., 2008). Copper has been previously shown to interact with PrP^C and to trigger

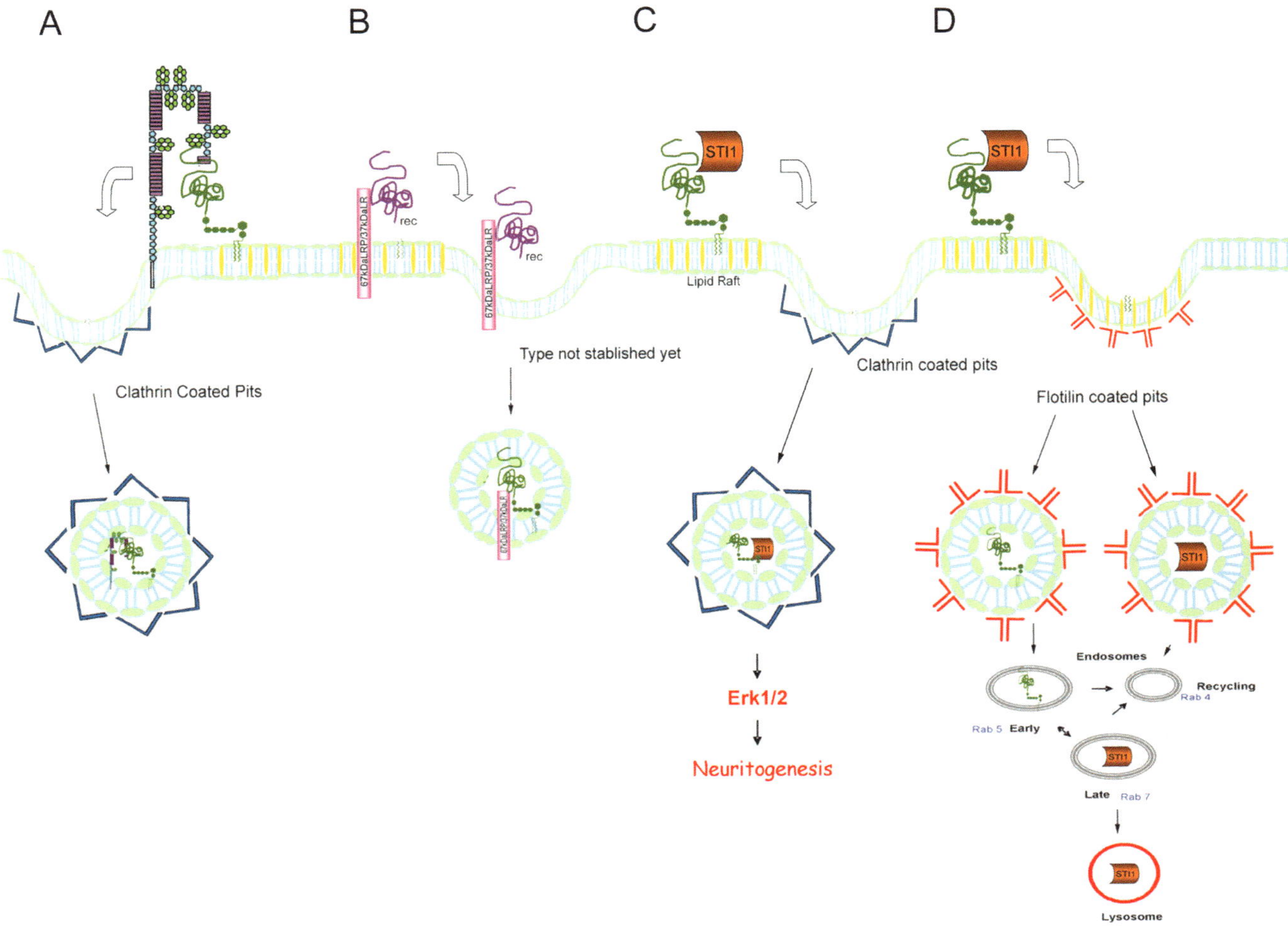

Figure 8. Distribution of PrP^C in specific membrane microdomains and endocytic vesicles mediates neurotrophic functions. A) LRP interacts with PrP^C located in lipid rafts and triggers the latter's internalization. B) 37LRP/67LR promotes the internalization of recombinant PrP^C through an unknown mechanism. C) STI1 induces PrP^C internalization by clathrin coated pits. ERK1/2 activation depends of PrP^C internalization. D) PrP^C and STI1 were also found in flotillin/reggie1 vesicles, recycling vesicles and lysosomes.

its endocytosis (Pauly and Harris, 1998; Lee et al., 2001; Perera and Hooper, 2001). Hemin (iron protoporphyrin IX) added to culture media can also evoke PrP^C internalization and recycling (Lee et al., 2007b). However, to the best of our knowledge, STI1 is the first ligand, which triggers cellular signaling by interacting with PrP^C that can also evoke sequestration of this protein. Internalization of cell surface receptors is usually considered a mechanism for modulation of cellular responses (Ferguson, 2001), suggesting the possibility that endocytosis of PrP^C triggered by STI1 may be involved in cellular signaling.

This possibility received initial support when it was found that treatments known to decrease endocytic activity in cells diminished STI1 mediated signaling (Americo et al., 2007). Recent work showed that a PrP^C mutant (^23KKRPKP^28 to ^23KQHPSP^28), that fails to undergo clathrin-mediated endocytosis (Sunyach et al., 2003), can reconstitute PrP^C mediated activation of PKA in response to STI1, but not activation of ERK1/2 activation in the same condition (Caetano et al., 2008). The most parsimonious explanation for these results is that activation of ERK1/2, but not PKA triggerd by STI1-PrP^C interaction depends on the internalization of PrP^C into clathrin-coated vesicles (Figure 8C).

Interestingly, the activation of ERK1/2 induced by STI1 is fast and transient, lasting no more than a few minutes (Lopes et al., 2005; Caetano et al., 2008). This observation has led us to investigate if STI1 trafficking in cells could explain the short activation time of ERK1/2. Indeed, STI1 is also internalized by cells in a PrP^C independent-way. A fraction of STI1 appears to be internalized with PrP^C and, in the initial moments of internalization part of STI1 is found in clathrin-coated vesicles. However, shortly after internalization STI1 and PrP^C appear to be sorted to distinct intracellular paths, and upon prolonged chase they distribute to distinct intracellular compartments (Caetano et al., 2008). Remarkably, most internalized STI1 is eventually found in flotillin/reggie-positive vesicles (Figure 8C). We have speculated that the divergence in intracellular roads taken by the two proteins may function to attenuate ERK1/2 signalling (Caetano et al., 2008).

PrP^C was also found in flotillin/reggie-1 lipid rafts in non-neuronal cells, suggesting that, at least in certain circumstances PrP^C and STI1 could share the same internalization pathway. PrP^C was targeted to these domains through antibody cross-linking, leading to actin polymerization, ERK1/2 phosphorylation and a distinct brief Ca^{2+} signal (Stuermer et al., 2004). These results also imply that PrP^C sequestration and activity are coupled events. Reggie/flotilins are cytoplasmic proteins involved in the capping of raft derived vesicles, and can also be found in the nervous system. Their ablation induces morphological defects in the nervous system during embryogenesis (Stuermer and Plattner, 2005), and they are involved in the secretion of morphogens such as Hedgehog and Wnt in *Drosophila* (Katanaev et al., 2008). Since both PrP^C and STI1 can be secreted, the interaction of these proteins with reggie/flotillin-positive organelles may have important implications for the availability of secreted PrP^C or STI1 (Figure 8D).

The possibility that PrP^C endocytosis is a triggered event related to cellular signaling merits further investigation. For example, as demonstrated in Figure 6C, PrP^C at the cell surface has been shown recently to constitutive silence the NR2D receptor subunit in hippocampal neurons (Khosravani et al., 2008). This subunit, when incorporated to functional NMDA receptors generates currents with long decay kinetics, which may be responsible for increased sensitivity to excitotoxicity in PrP^C-null neurons (Khosravani et al., 2008). It could be envisioned that sequestration of cell surface PrP^C may be a mechanism to transiently increase NMDA currents in hippocampal neurons. Another potential site for regulation is the β-secretase pathway involved in APP processing that has been recently shown to be regulated by the presence of PrP^C in lipid rafts (Parkin et al., 2007). Future experiments aimed at dissecting signaling complexes involved in PrP^C function, should also target the mechanisms through which these complexes regulate and are regulated by cellular prion trafficking.

PrP^C in astrocytes promoting neuronal survival and differentiation

Astrocytes and neurons interact in various neurophysiological events and represent an integral unit in brain function. Astrocytes are key elements in both the production of neurotrophic molecules and controlling the levels of neurotoxic substances. They can also modulate the production of extracellular matrix that supports neuronal development and differentiation (Theodosis et al., 2008).

The functions of PrP^C in astrocytes are largely unknown. It was shown that PrP^C associates with both the GluR2 subunit of AMPA receptors and 2α/β2 Na^+/K^+ATPase. The latter binds to basigin, a cell adhesion molecule (Kleene et al., 2007), which is tightly associated with the lactate monocarboxylate transporter 1 (MCT1). In the absence of glutamate, the 2α/β2 Na^+/K^+ATPase is less active, and MCT1 does not transport lactate. Conversely, when glutamate is present, MCT1 is active, leading to either increased release or uptake of lactate, depending on the concentration of the latter inside and outside of the astrocyte. Intense synaptic activity implies an increased energy demand, and lactate provided by astrocytes may be a necessary alternative substrate for balancing the energetics of neuronal activity. Deletion of PrP^C is followed by inactivation of 2α/β2 Na^+/K^+ATPase, and an interruption on the control of MCT1, which becomes permanently active even in the absence of glutamate. It was suggested that the deletion of PrP^C may result in increased levels of extracellular lactate, and lactate-induced acidosis leading to neuronal damage (Kleene et al., 2007). The PrP^C-dependent protein complex involved in lactate transport may therefore play an important role upon synaptic function (Figure 9A).

The role of PrP^C in neurotrophic interactions between astrocytes and neurons has also been addressed (Lima et al., 2007). Prominent neurite outgrowth was observed when wild-type neurons were co-cultured with wild-type astrocytes, while poor differentiation was present when both cells were derived from PrP^C-null mice. It is possible that PrP^C in astrocytes contributes to neurite outgrowth by trans-activating a ligand such as NCAM in neurons (Figure 9B). In addition, astrocytes produce and secrete laminin, among other ECM proteins. Laminin secreted and deposited at the ECM by wild-type astrocytes showed a fibrilar organization (Figure 9B), while that secreted from PrP^C-null astrocytes has a punctate pattern (Figure 9C). The pattern of laminin matrix has been shown to influence neurite outgrowth (Freire et al., 2002), either through modulation of its binding to PrP^C (Graner et al., 2000a) or to other cell surface receptors (Denda and Reichardt, 2007; Nelson et al., 2008).

Astrocytes also release soluble neurotrophic factors and proteins such as STI1 and PrP^C (Lima et al., 2007) and these proteins can bind to neuronal PrP^C and NCAM respectively, mediating neuronal survival and differentiation (Lima et al., 2007). Neuronal survival was higher when cells where treated with conditioned medium from wild-type astrocytes, as compared with medium from PrP^C-null astrocytes. These data clearly show that PrP^C in astrocytes is critical to sustain cell-cell interaction, organization of extracellular matrix and secretion of soluble factors, thus helping to maintain a healthy environment for neuronal survival and differentiation (Figures 9B and C).

Therefore, PrP^C not only works as a neurotrophic factor for neuronal cells, but also modulates astrocyte performance on neuronal survival and differentiation.

PrP^C as a therapeutic target for degenerative illnesses

Given that PrP^C interacts with several classes of molecules triggering neurotrophic functions, and the high expression of PrP^C in the nervous system, it is conceivable that the modulation of PrP^C activity and its interaction with specific ligands may constitute therapeutic targets not only for prion diseases, but also for other chronic degenerative diseases.

The pathogenesis of prion diseases can be explained, at least in part, by "loss-of-function" mechanism, which consists in the depletion of the neuroprotective function of PrP^C, when converted to PrP^{Sc}. However, this mecanism is still debatable, due to some noteworthy results. For example, the deletion of neuronal PrP^C in adult mice causes changes in some hippocampal properties, but no neurodegeneration (Mallucci et al., 2002). Spongiosis, neuron loss, cognitive and behavioral deficits, as well as impaired neurophysiological function, were reverted upon deletion of PrP^C in early stages of the disease in mice (Mallucci et al., 2003; Mallucci et al., 2007). Remarkably, these results were reproduced when the expression of PrP^C was inhibited by a lentiviral vector containing an anti-PrP^C shRNA in prion infected mice (White et al., 2008). This treatment also significantly prolonged the survival of those animals (White et al., 2008). Thus, despite limitations such as treatment in early stages of prion disease, the need

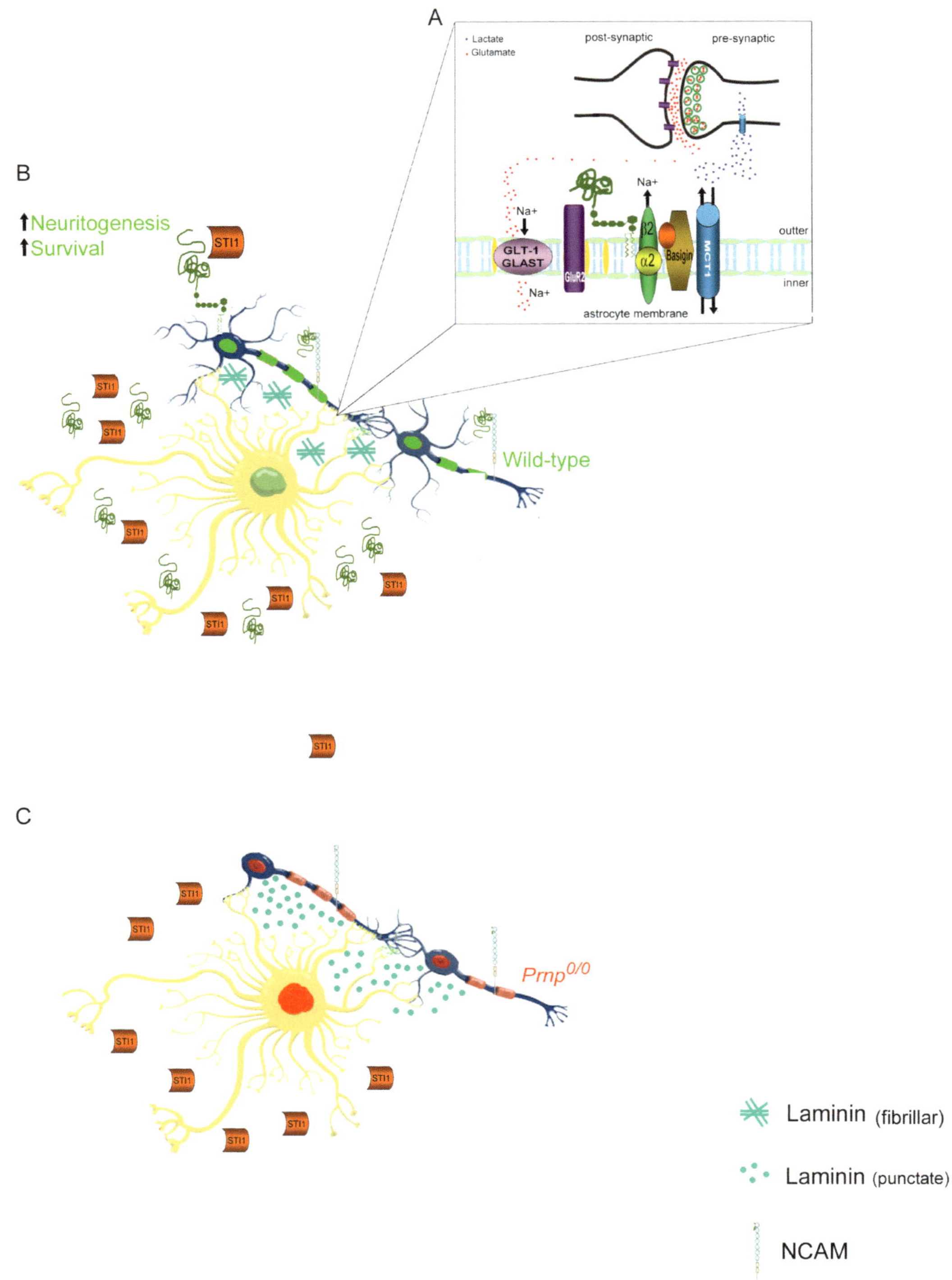

Figure 9. PrP^C modulates neurotrophic properties in astrocytes. A) Glutamate released in synapses is taken up by astrocytes via glial glutamate transporters (GLAST/GLT-1). Na^+ dependent transport of glutamate causes a massive Na^+ influx into the cell, leading to the activation of the Na^+/K^+ ATPase. In order to maintain a high glycolitic flux, lactate is released from astrocytes into the extracellular space via the monocarboxylate trasporter (MCT-1). Lactate is taken up by neurons. PrP^C associates with a protein-complex involved in lactate transport in the astrocytic membrane. PrP^C can directly interact with both, the GluR2 subunit of AMPA receptors and $2\alpha/\beta2$ Na^+/K^+ATPase. The latter binds a cell adhesion molecule, basigin, required as an auxiliary protein to maintain the catalytic activity of the lactate receptors. The lack of PrP^C reduces the $2\alpha/\beta2$ Na^+/K^+ATPase pump activity and activates basigin/MCT1associate lactate transport. B and C) PrP^C and STI1 as neurotrophic factors in neuron-glia interaction. A prominent neuritogenesis is observed when wild-type neurons are co-cultured with wild-type astrocytes (B) neurite outgrowth is impaired when both cell types are derived from PrP^C-null mice (C). Laminin secreted and deposited at the ECM by wild-type astrocytes has a fibrillar organization and promotes more intense neurite outgrowth than the puctated laminin secreted from PrP^C-null astrocytes. Astrocytes also release STI1 and PrP^C, which induce neuronal survival and differentiation.

for transport across the blood-brain barrier, and off-target effects, anti-PrP^C siRNA seems to be a promising approach to treat prion diseases. Notwithstanding, its effect regarding PrP^C loss-of-function in humans need to be considered. Still, short lived side effects may be acceptable, due to the devastating characteristics of prion diseases.

The use of anti-PrP^C antibodies to block prion infection *in vivo* has also been addressed, but some of the data are contraditory, particulary regarding the toxicity of the treatment. The expression of a transgenic anti-PrP^C IgMa µ chain prevented pathogenesis of prions directly inoculated into brain or spleen (Heppner et al., 2001). In addition, peripheral infection was markedly reduced by treatment with anti-PrP^C antibodies (White et al., 2003). Extracerebrally infected CD-1 mice also presented effective suppression of PrP^{Sc} replication in the lymphoreticular system, lengthening of the incubation period and a significant decrease in CNS pathology after treatment with anti-PrP^C antibodies (Sadowski et al., 2009). However, important concerns about intracerebral treatment with antibodies to PrP^C were raised (Tayebi and Hawke, 2006). The cross-linking of PrP^C *in vivo* with specific monoclonal antibodies triggers rapid and extensive apoptosis in hippocampal and cerebellar neurons (Solforosi et al., 2004). Prolonged treatment with PrP^C antibodies also causes extensive neuronal loss, strong astrogliosis and microglial activation, even when IgG Fab fragments were applied (Lefebvre-Roque et al., 2007). In addition, this treatment was not protective against the development of BSE in transgenic mice overexpressing PrP^C (Lefebvre-Roque et al., 2007).

Blockade of PrP^C-PrP^{Sc} conversion without altering PrP^C function is also an interesting alternative for prion diseases. This has been approached through the targeting of physiological ligands already characterized as inhibitors of PrP^C conversion into PrP^{Sc}. A trans-dominant negative 37LRP/67LR mutant reportedly had an inhibitory effect on PrP^C-PrP^{Sc} conversion *in vitro* (Vana and Weiss, 2006). Preclinical studies demonstrated that passive immunotransfer of single chain Fv antibodies directed against the 37LRP/67LR reduced PrP^{Sc} levels in the spleen of prion infected mice (Zuber et al., 2008). Stereotatic intracerebral microinjection of adeno-associated virus (AAV) vectors expressing single chain Fv antibodies against this receptor leads to a reduction of peripheral PrP^{Sc}, however without a significant lenghtening of incubation times and survival (Zuber et al., 2008). In this line, other candidate targets may be glycosaminoglycans (GAGs), which have both high affinity for PrP^C and demonstrated anti-scrapie activities (Caughey et al., 1994; Warner et al., 2002; Schonberger et al., 2003; Kocisko et al., 2006; Caughey et al., 2006; Silva et al., 2008). Further studies are required to test how the modulation of these molecules for the treatment of prion diseases may affect physiological functions of PrP^C.

PrP^C may also be approached as a candidate neuroprotective agent in Alzheimer's disease, due to its effect of downmodulating the activity of β-secretase, therefore decreasing the production of the toxic amyloid-β (Aβ) peptide, the main component of senile plaques (Parkin et al., 2007). On the other hand, PrP^C was also found to be the major receptor for Aβ oligomers in neurons and their interaction is important for the effects of Aβ on synaptic disfunction (Lauren et al., 2009). The possible use of PrP^C as a therapeutic target in Alzheimer diseases has been explored (Parkin et al., 2007; Lauren et al., 2009), and the development of reagents that increase β-secretase inhibition by PrP^C and impair PrP^C-Aβ oligomers binding would be desirable.

Although exogenous GAGs may constitute lead compounds for the development of therapeutic methods for prion diseases (Schonberger et al., 2003), the same can be detrimental for Alzheimer´s disease. Whereas PrP^C interacts with BACE1 therefore inhibiting its β-secretase activity, the interaction between PrP^C and BACE1 involves GAGs, and the exogenous GAGs appear to disrupt this interaction, thus restoring the production of Aβ peptides (Parkin et al., 2007). Still, exploring PrP^C–GAGs interactions would potentially generate new and improved therapeutic strategies for both prion and Alzheimer´s diseases.

Both the differentiation of dorsal root ganglia (Hajj et al., 2007) as well as motor neuron survival (Barbeito A., Hajj G, Martins VR, Barbeito L unpublished results) can be modulated by PrP^C. Although previous results suggested downregulation of PrP^C in a transgenic mouse model for amyotrophic lateral sclerosis (ALS) (Dupuis et al., 2002) our data demonstrated that PrP^C expression in tissues from similar animals are comparable to their normal counterpart (Barbeito A., Hajj G, Martins VR, Barbeito L unpublished results). Therefore, protective effects of PrP^C in motor neurons can also become a therapeutic target for ALS, and this hypothesis deserves to be explored.

It has been reported that PrP^C depletion causes a higher aggregation of huntingtin (Htt), similar to what is observed when cells are treated with a proteasome inhibitor. Conversely, overexpression of PrP^C in cells expressing Htt caused a decrease in both the number of cells with Htt granules and the number of apoptotic cells, indicating a protective activity of PrP^C against pathological protein aggregation (Lee et al., 2007a). Histone deacetylase (HDAC) inhibitors have been proposed as a novel therapeutic approach for the Huntington's disease (Sadri-Vakili and Cha, 2006) and PrP^C is upregulated by such inhibitors (Cabral et al., 2002). Altogether, these data suggest that PrP^C should be considered in attempts to develop new therapeutic approaches for Huntington´s disease.

The protective function of PrP^C extends to other neurological disorders such as epilepsy. Two independent studies demonstrated that PrP^C-null mice are more susceptible to seizures induced by convulsant agents (Walz et al., 1999; Rangel et al., 2007). After injection of kainate or pentylenetetrazol, these animals presented higher grade seizures reaching elevated mortality and increased neuronal death when compared with control mice (Walz et al., 1999; Rangel et al., 2007). Moreover, several brain regions of PrP^C-null mice showed increased neuronal death upon injections of kainate (Rangel et al., 2007). The absence of PrP^C decreased the activity of GABAergic receptors (Collinge et al., 1994) and increased the expression and activity of NMDA receptors (Maglio et al., 2004; Khosravani et al., 2008). Thus, the neuroprotective function of PrP^C against seizures may be a consequence of its ability to balance inhibitory and excitatory synaptic transmission (Khosravani et al., 2008).

The expression of PrP^C is also important for the maintenance of the circadian rhythm and sleep homeostasis. Animals that do not express PrP^C presented longer period of activity in free-running conditions compared to normal mice and this phenotype was rescued with re-expression of PrP^C (Tobler et al., 1996). Moreover, the lack of PrP^C expression increased the sleep fragmentation indicating that PrP^C is required for sleep continuity (Tobler et al., 1996; Tobler et al., 1997). Consistently, patients with prion diseases also present altered sleep patterns that can extend to the case of patients carrying a mutation Asp178 Asn that leads to fatal familial insomnia (Landolt et al., 2006; Provini et al., 2008). Since sleep is a physiological process tightly controlled by various neurotransmitter systems, these data corroborate the view that PrP^C can be a target to modulate synaptic transmission and for the treatment of sleep disorders.

Finally, PrP^C can also be a target for the treatment of acute injuries. The induction of ischemic stroke in the animal model generates complex pathophysiological conditions in the brain that can culminate with cell death. This experimental approach was used by several groups to evaluate the neuroprotective role of PrP^C *in vivo*. All findings unanimously demonstrated that ischemic lesions increased the local expression of PrP^C and that PrP^C-null mice were more vulnerable to the ischemic stroke presenting larger brain injury (McLennan et al., 2004; Weise et al., 2004; Shyu et al., 2005; Weise et al., 2006). The up-regulation of PrP^C due to the ischemic stroke was also observed in human post-mortem brain tissue. Pyramidal neurons, inflammatory cells and micro vessels of the peri-infarcted area expressed higher level of PrP^C compared to contralateral tissues (Mitsios et al., 2007). Moreover, a higher level of PrP^C was also present in the plasma (Mitsios et al., 2007). The up-regulation of PrP^C expression in the ischemic tissue may represent a protective mechanism against the injury and might conceivably be explored therapeuticaly.

Conclusion and future directions

Studies of neurotrophic signal transduction have evolved from the search for simple pathways controlling cell survival to complex, non-linear systems, including interactions of classic neurotrophic factors (Sharma, 2007; Sanford et al., 2008; Hossain et al., 2008), cross-talk and transactivation among differing neuroactive molecules and membrane receptors (Lee et al., 2002; Wiese et al., 2007; Jeanneteau et al., 2008; Iwakura et al., 2008), selective, and even opposite, responses either of individual receptors or to single neurotrophic molecules (Chao et al., 2002; Iwakura et al., 2008), modulation of receptor expression and exposure (Nagappan and Lu, 2005), as well as subcellular traffic-dependent signaling (Howe and Mobley, 2005; Ibanez, 2007; Cosker et al., 2008).

Scaffolding proteins provide for both spatial coordination and kinetics of molecular interactions, which are critical for the selective activation of signaling cascades within the intracellular environment (Pawson and Scott, 1997; Bhattacharyya et al., 2006; Murrin and Talbot, 2007; Jarnaess and Tasken, 2007; Bezman and Koretzky, 2007; Dohlman, 2008). A number of intracellular scaffolding proteins are required for signaling downstream of membrane receptors and for interaction of known neurotrophic molecules (Nakaoka et al., 2003; Hisata et al., 2007; Iida et al., 2007; Kurisaki et al., 2008).

We have proposed that the main function of the prion protein is to serve as a dynamic cell surface scaffolding protein, based on its ability to bind a large number of ligands, and to travel across boundaries of distinct plasma membrane domains, as well as to move bidirectionally between the cell surface and intracellular endocytic compartments. These properties allow for a role of PrP^C upon both the selection and the kinetics of extracellular and cell surface components of signaling modules, as well as the recruiting of transmembrane components able to convey extracellular signals through the plasma membrane. Such properties are likely to underlie both the varied and often contradictory functions attributed to PrP^C, and the multiple phenotypes that follow modulation of the expression of PrP^C (Linden et al., 2008).

The current review underscores the importance of PrP^C in the coordination of signaling related to neuronal survival and differentiation. The ability of PrP^C to mediate the assembly of multi-component complexes at the cell surface is likely the basis for its neurotrophic properties, and may be relevant for loss-of-function components of prion diseases (Samaia and Brentani, 1998), in particular those associated with neuronal death. In addition, the wide range of molecular interactions described for PrP^C suggests that this protein may be relevant for not only the spongiform encephalophaties, but also as an ancillary component of the pathogenesis of other neurodegenerative diseases, and therefore amenable to therapeutic targeting. Further studies should be directed at the unraveling of additional signaling complexes scaffolded by PrP^C, as well as to the role of these molecular assemblies upon both physiological and pathological properties of both the nervous system and other organs (Linden et al., 2008).

Acknowledgements

Research done in the authors' laboratories, included in this review, was supported by grants and fellowships from FAPESP, CNPq, CAPES, FAPERJ, PRONEX, The Millenium Institutes – Gene Therapy Network/MCT and Toxins/MCT, FINEP, FAPEMIG and NIH-Fogarty Center. R.L. and M.A.M.P. are former fellows of the John Simon Guggenheim Foundation. Fellowships from FAPESP to F.H.B.P, M.H.L. and G.N.M.H. are gratefully acknowledged V.R.M is an International Research Scholar of the Howard Hughes Medical Institute.

References

Alfa CM, Sunyach C, Slack BE, Fisher A, Vincent B, and Checler F (2007) M1 and M3 muscarinic receptors control physiological processing of cellular prion by modulating ADAM17 phosphorylation and activity. J Neurosci, 27, 4083-4092.

Americo TA, Chiarini LB, and Linden R (2007) Signaling induced by hop/STI-1 depends on endocytosis. Biochem Biophys Res Commun, 358, 620-625.

Azmitia EC (2002) Cajal's hypotheses on neurobiones and neurotropic factor match properties of microtubules and S-100 beta. Prog Brain Res, 136, 87-100.

Azzalin A, Ferrara V, Arias A, Cerri S, Avella D, Pisu MB, Nano R, Bernocchi G, Ferretti L, and Comincini S (2006) Interaction between the cellular prion (PrPC) and the 2P domain K+ channel TREK-1 protein. Biochem Biophys Res Commun, 346, 108-115.

Bainbridge J and Walker KB (2005) The normal cellular form of prion protein modulates T cell responses. Immunol Lett, 96, 147-150.

Baumann F, Tolnay M, Brabeck C, Pahnke J, Kloz U, Niemann HH, Heikenwalder M, Rulicke T, Burkle A, and Aguzzi A (2007) Lethal recessive myelin toxicity of prion protein lacking its central domain. EMBO J, 26, 538-547.

Beringue V, Mallinson G, Kaisar M, Tayebi M, Sattar Z, Jackson G, Anstee D, Collinge J, and Hawke S (2003) Regional heterogeneity of cellular prion protein isoforms in the mouse brain. Brain, 126, 2065-2073.

Bezman N and Koretzky GA (2007) Compartmentalization of ITAM and integrin signaling by adapter molecules. Immunol Rev, 218, 9-28.

Bhattacharyya RP, Remenyi A, Yeh BJ, and Lim WA (2006) Domains, motifs, and scaffolds: the role of modular interactions in the evolution and wiring of cell signaling circuits. Annu Rev Biochem, 75, 655-680.

Blochl A and Blochl R (2007) A cell-biological model of p75NTR signaling. J Neurochem, 102, 289-305.

Borchelt DR, Rogers M, Stahl N, Telling G, and Prusiner SB (1993) Release of the cellular prion protein from cultured cells after loss of its glycoinositol phospholipid anchor. Glycobiology, 3, 319-329.

Brini M, Miuzzo M, Pierobon N, Negro A, and Sorgato MC (2005) The prion protein and its paralogue Doppel affect calcium signaling in Chinese hamster ovary cells. Mol Biol Cell, 16, 2799-2808.

Bueler H, Fischer M, Lang Y, Bluethmann H, Lipp HP, DeArmond SJ, Prusiner SB, Aguet M, and Weissmann C (1992) Normal development and behaviour of mice lacking the neuronal cell- surface PrP protein. Nature, 356, 577-582.

Cabral AL, Lee KS, and Martins VR (2002) Regulation of the cellular prion protein gene expression depends on chromatin conformation. J Biol Chem, 277, 5675-5682.

Caetano FA, Lopes MH, Hajj GN, Machado CF, Pinto AC, Magalhaes AC, Vieira MP, Americo TA, Massensini AR, Priola SA, Vorberg I, Gomez MV, Linden R, Prado VF, Martins VR, and Prado MA (2008) Endocytosis of prion protein is required for ERK1/2 signaling induced by stress-inducible protein 1. J Neurosci, 28, 6691-6702.

Calixto JB, Medeiros R, Fernandes ES, Ferreira J, Cabrini DA, and Campos MM (2004) Kinin B1 receptors: key G-protein-coupled receptors and their role in inflammatory and painful processes. Br J Pharmacol, 143, 803-818.

Calzolai L, Lysek DA, Perez DR, Guntert P, and Wuthrich K (2005) Prion protein NMR structures of chickens, turtles, and frogs. Proc Natl Acad Sci U S A, 102, 651-655.

Caughey B, Brown K, Raymond GJ, Katzenstein GE, and Thresher W (1994) Binding of the protease-sensitive form of PrP (prion protein) to sulfated glycosaminoglycan and congo red [corrected]. J Virol, 68, 2135-2141.

Caughey B, Caughey WS, Kocisko DA, Lee KS, Silveira JR, and Morrey JD (2006) Prions and transmissible spongiform encephalopathy (TSE) chemotherapeutics: A common mechanism for anti-TSE compounds? Acc Chem Res, 39, 646-653.

Chao MV (2003) Neurotrophins and their receptors: a convergence point for many signalling pathways. Nat Rev Neurosci, 4, 299-309.

Chao MV and Bothwell M (2002) Neurotrophins: to cleave or not to cleave. Neuron, 33, 9-12.

Chen S, Mange A, Dong L, Lehmann S, and Schachner M (2003) Prion protein as trans-interacting partner for neurons is involved in neurite outgrowth and neuronal survival. Mol Cell Neurosci, 22, 227-233.

Chen SG, Teplow DB, Parchi P, Teller JK, Gambetti P, and utilio-Gambetti L (1995) Truncated forms of the human prion protein in normal brain and in prion diseases. J Biol Chem, 270, 19173-19180.

Chiarini LB, Freitas AR, Zanata SM, Brentani RR, Martins VR, and Linden R (2002) Cellular prion protein transduces neuroprotective signals. EMBO J, 21, 3317-3326.

Cisse MA, Sunyach C, Lefranc-Jullien S, Postina R, Vincent B, and Checler F (2005) The disintegrin ADAM9 indirectly contributes to the physiological processing of cellular prion by modulating ADAM10 activity. J Biol Chem, 280, 40624-40631.

Clegg DO, Wingerd KL, Hikita ST, and Tolhurst EC (2003) Integrins in the development, function and dysfunction of the nervous system. Front Biosci, 8, d723-d750.

Coitinho AS, Freitas AR, Lopes MH, Hajj GN, Roesler R, Walz R, Rossato JI, Cammarota M, Izquierdo I, Martins VR, and Brentani RR (2006) The interaction between prion protein and laminin modulates memory consolidation. Eur J Neurosci, 24, 3255-3264.

Coitinho AS, Lopes MH, Hajj GN, Rossato JI, Freitas AR, Castro CC, Cammarota M, Brentani RR, Izquierdo I, and Martins VR (2007) Short-term memory formation and long-term memory consolidation are enhanced by cellular prion association to stress-inducible protein 1. Neurobiol Dis, 26, 282-290.

Coitinho AS, Roesler R, Martins VR, Brentani RR, and Izquierdo I (2003) Cellular prion protein ablation impairs behavior as a function of age. Neuroreport, 14, 1375-1379.

Colling SB, Collinge J, and Jefferys JG (1996) Hippocampal slices from prion protein null mice: disrupted Ca(2+)-activated K+ currents. Neurosci Lett, 209, 49-52.

Collinge J, Whittington MA, Sidle KC, Smith CJ, Palmer MS, Clarke AR, and Jefferys JG (1994) Prion protein is necessary for normal synaptic function. Nature, 370, 295-297.

Colognato H, ffrench-Constant C, and Feltri ML (2005) Human diseases reveal novel roles for neural laminins. Trends Neurosci, 28, 480-486.

Colognato H and Yurchenco PD (2000) Form and function: the laminin family of heterotrimers. Dev Dyn, 218, 213-234.

Conese M, Nykjaer A, Petersen CM, Cremona O, Pardi R, Andreasen PA, Gliemann J, Christensen EI, and Blasi F (1995) alpha-2 Macroglobulin receptor/Ldl receptor-related protein(Lrp)-dependent internalization of the urokinase receptor. J Cell Biol, 131, 1609-1622.

Cosker KE, Courchesne SL, and Segal RA (2008) Action in the axon: generation and transport of signaling endosomes. Curr Opin Neurobiol. 18, 270-275.

Coulpier M, Messiaen S, Boucreaux D, and Eloit M (2006) Axotomy-induced motoneuron death is delayed in mice overexpressing PrPc. Neuroscience, 141, 1827-1834.

Criado JR, Sanchez-Alavez M, Conti B, Giacchino JL, Wills DN, Henriksen SJ, Race R, Manson JC, Chesebro B, and Oldstone MB (2005) Mice devoid of prion protein have cognitive deficits that are rescued by reconstitution of PrP in neurons. Neurobiol Dis, 19, 255-265.

Czekay RP, Kuemmel TA, Orlando RA, and Farquhar MG (2001) Direct binding of occupied urokinase receptor (uPAR) to LDL receptor-related protein is required for endocytosis of uPAR and regulation of cell surface urokinase activity. Mol Biol Cell, 12, 1467-1479.

Danielisova V, Gottlieb M, Nemethova M, and Burda J (2008) Effects of bradykinin postconditioning on endogenous antioxidant enzyme activity after transient forebrain ischemia in rat. Neurochem Res, 33, 1057-1064.

Datson NA, Morsink MC, Meijer OC, and de Kloet ER (2008) Central corticosteroid actions: Search for gene targets. Eur J Pharmacol, 583, 272-289.

Davies AM (1996) The neurotrophic hypothesis: where does it stand? Philos Trans R Soc Lond B Biol Sci, 351, 389-394.

de Almeida CJ, Chiarini LB, da Silva JP, PM ES, Martins MA, and Linden R (2005) The cellular prion protein modulates phagocytosis and inflammatory response. J Leukoc Biol, 77, 238-246.

De AA and Georges-Labouesse E (2000) Integrin and ECM functions: roles in vertebrate development. Trends Genet, 16, 389-395.

DeArmond SJ, Qiu Y, Sanchez H, Spilman PR, Ninchak-Casey A, Alonso D, and Daggett V (1999) PrPc glycoform heterogeneity as a function of brain region: implications for selective targeting of neurons by prion strains. J Neuropathol Exp Neurol, 58, 1000-1009.

DeArmond SJ, Sanchez H, Yehiely F, Qiu Y, Ninchak-Casey A, Daggett V, Camerino AP, Cayetano J, Rogers M, Groth D, Torchia M, Tremblay P, Scott MR, Cohen FE, and Prusiner SB (1997) Selective neuronal targeting in prion disease. Neuron, 19, 1337-1348.

Denda S and Reichardt LF (2007) Studies on integrins in the nervous system. Methods Enzymol, 426, 203-221.

Ditlevsen DK, Povlsen GK, Berezin V, and Bock E (2008) NCAM-induced intracellular signaling revisited. J Neurosci Res, 86, 727-743.

Dohlman HG (2008) A scaffold makes the switch. Sci Signal, 1, e46.

Dupuis L, Mbebi C, Gonzalez de Aguilar JL, Rene F, Muller A, de TM, and Loeffler JP (2002) Loss of prion protein in a transgenic model of amyotrophic lateral sclerosis. Mol Cell Neurosci, 19, 216-224.

Eberl H, Tittmann P, and Glockshuber R (2004) Characterization of recombinant, membrane-attached full-length prion protein. J Biol Chem, 279, 25058-25065.

Fagan KA, Graf RA, Tolman S, Schaack J, and Cooper DM (2000) Regulation of a Ca2+-sensitive adenylyl cyclase in an excitable cell. Role of voltage-gated versus capacitative Ca2+ entry. J Biol Chem, 275, 40187-40194.

Ferguson SS (2001) Evolving concepts in G protein-coupled receptor endocytosis: the role in receptor desensitization and signaling. Pharmacol Rev, 53, 1-24.

Fevrier B, Vilette D, Archer F, Loew D, Faigle W, Vidal M, Laude H, and Raposo G (2004) Cells release prions in association with exosomes. Proc Natl Acad Sci U S A, 101, 9683-9688.

Fink M, Duprat F, Lesage F, Reyes R, Romey G, Heurteaux C, and Lazdunski M (1996) Cloning, functional expression and brain localization of a novel unconventional outward rectifier K+ channel. EMBO J, 15, 6854-6862.

Fivaz M, Vilbois F, Thurnheer S, Pasquali C, Abrami L, Bickel PE, Parton RG, and van der Goot FG (2002) Differential sorting and fate of endocytosed GPI-anchored proteins. EMBO J, 21, 3989-4000.

Franks NP and Honore E (2004) The TREK K2P channels and their role in general anaesthesia and neuroprotection. Trends Pharmacol Sci, 25, 601-608.

Freire E, Gomes FC, Linden R, Neto VM, and Coelho-Sampaio T (2002) Structure of laminin substrate modulates cellular signaling for neuritogenesis. J Cell Sci, 115, 4867-4876.

Fuhrmann M, Bittner T, Mitteregger G, Haider N, Moosmang S, Kretzschmar H, and Herms J (2006) Loss of the cellular prion protein affects the Ca2+ homeostasis in hippocampal CA1 neurons. J Neurochem, 98, 1876-1885.

Gauczynski S, Peyrin JM, Haik S, Leucht C, Hundt C, Rieger R, Krasemann S, Deslys JP, Dormont D, Lasmezas CI, and Weiss S (2001) The 37-kDa/67-kDa laminin receptor acts as the cell-surface receptor for the cellular prion protein. EMBO J, 20, 5863-5875.

Giacomello M, Drago I, Pizzo P, and Pozzan T (2007) Mitochondrial Ca2+ as a key regulator of cell life and death. Cell Death Differ, 14, 1267-1274.

Gorodinsky A and Harris DA (1995) Glycolipid-anchored proteins in neuroblastoma cells form detergent-resistant complexes without caveolin. J Cell Biol, 129, 619-627.

Graner E, Mercadante AF, Zanata SM, Forlenza OV, Cabral AL, Veiga SS, Juliano MA, Roesler R, Walz R, Minetti A, Izquierdo I, Martins VR, and Brentani, RR (2000a) Cellular prion protein binds laminin and mediates neuritogenesis. Brain Res Mol Brain Res, 76, 85-92.

Graner E, Mercadante AF, Zanata SM, Martins VR, Jay DG, and Brentani RR (2000b) Laminin-induced PC-12 cell differentiation is inhibited following laser inactivation of cellular prion protein. FEBS Lett, 482, 257-260.

Grimpe B and Silver J (2002) The extracellular matrix in axon regeneration. Prog Brain Res, 137, 333-349.

Hajj GN, Lopes MH, Mercadante AF, Veiga SS, da Silveira RB, Santos TG, Ribeiro KC, Juliano MA, Jacchieri SG, Zanata SM, and Martins VR (2007) Cellular prion protein interaction with vitronectin supports axonal growth and is compensated by integrins. J Cell Sci, 120, 1915-1926.

Haraguchi T, Fisher S, Olofsson S, Endo T, Groth D, Tarentino A, Borchelt DR, Teplow D, Hood L, Burlingame A, and . (1989) Asparagine-linked glycosylation of the scrapie and cellular prion proteins. Arch Biochem Biophys, 274, 1-13.

Harris DA, Huber MT, van DP, Shyng SL, Chait BT, and Wang R (1993) Processing of a cellular prion protein: identification of N- and C-terminal cleavage sites. Biochemistry, 32, 1009-1016.

Hegde RS, Mastrianni JA, Scott MR, DeFea KA, Tremblay P, Torchia M, DeArmond SJ, Prusiner SB, and Lingappa VR (1998) A transmembrane form of the prion protein in neurodegenerative disease. Science, 279, 827-834.

Heppner FL, Musahl C, Arrighi I, Klein MA, Rulicke T, Oesch B, Zinkernagel RM, Kalinke U, and Aguzzi A (2001) Prevention of scrapie pathogenesis by transgenic expression of anti-prion protein antibodies. Science, 294, 178-182.

Herms JW, Korte S, Gall S, Schneider I, Dunker S, and Kretzschmar HA (2000) Altered intracellular calcium homeostasis in cerebellar granule cells of prion protein-deficient mice. J Neurochem, 75, 1487-1492.

Herms JW, Tings T, Dunker S, and Kretzschmar HA (2001) Prion protein affects Ca2+-activated K+ currents in cerebellar purkinje cells. Neurobiol Dis, 8, 324-330.

Hicks MR, Gill AC, Bath IK, Rullay AK, Sylvester ID, Crout DH, and Pinheiro TJ (2006) Synthesis and structural characterization of a mimetic membrane-anchored prion protein. FEBS J, 273, 1285-1299.

Hisata S, Sakisaka T, Baba T, Yamada T, Aoki K, Matsuda M, and Takai Y (2007) Rap1-PDZ-GEF1 interacts with a neurotrophin receptor at late endosomes, leading to sustained activation of Rap1 and ERK and neurite outgrowth. J Cell Biol, 178, 843-860.

Hornemann S, Schorn C, and Wuthrich K (2004) NMR structure of the bovine prion protein isolated from healthy calf brains. EMBO Rep, 5, 1159-1164.

Hossain WA, D'Sa C, and Morest DK (2008) Interactive roles of fibroblast growth factor 2 and neurotrophin 3 in the sequence of migration, process outgrowth, and axonal differentiation of mouse cochlear ganglion cells. J Neurosci Res, 86, 2376-2391.

Howe CL and Mobley WC (2005) Long-distance retrograde neurotrophic signaling. Curr Opin Neurobiol, 15, 40-48.

Huang EJ and Reichardt LF (2003) Trk receptors: roles in neuronal signal transduction. Annu Rev Biochem, 72, 609-642.

Hundt C, Peyrin JM, Haik S, Gauczynski S, Leucht C, Rieger R, Riley ML, Deslys JP, Dormont D, Lasmezas CI, and Weiss S (2001) Identification of interaction domains of the prion protein with its 37-kDa/67-kDa laminin receptor. EMBO J, 20, 5876-5886.

Ibanez CF (2007) Message in a bottle: long-range retrograde signaling in the nervous system. Trends Cell Biol, 17, 519-528.

Iida J, Ishizaki H, Okamoto-Tanaka M, Kawata A, Sumita K, Ohgake S, Sato Y, Yorifuji H, Nukina N, Ohashi K, Mizuno K, Tsutsumi T, Mizoguchi A, Miyoshi J, Takai Y, and Hata Y (2007) Synaptic scaffolding molecule alpha is a scaffold to mediate N-methyl-D-aspartate receptor-dependent RhoA activation in dendrites. Mol Cell Biol, 27, 4388-4405.

Iwakura Y, Nawa H, Sora I, and Chao MV (2008) Dopamine D1 receptor-induced signaling through TrkB receptors in striatal neurons. J Biol Chem, 283, 1579915806.

Jarnaess E and Tasken K (2007) Spatiotemporal control of cAMP signalling processes by anchored signalling complexes. Biochem Soc Trans, 35, 931-937.

Jeanneteau F, Garabedian MJ, and Chao MV (2008) Activation of Trk neurotrophin receptors by glucocorticoids provides a neuroprotective effect. Proc Natl Acad Sci U S A, 105, 4862-4867.

Jodoin J, Laroche-Pierre S, Goodyer CG, and Leblanc AC (2007) Defective retrotranslocation causes loss of anti-Bax function in human familial prion protein mutants. J Neurosci, 27, 5081-5091.

Kalb R (2005) The protean actions of neurotrophins and their receptors on the life and death of neurons. Trends Neurosci, 28, 5-11.

Kanaani J, Prusiner SB, Diacovo J, Baekkeskov S, and Legname G (2005) Recombinant prion protein induces rapid polarization and development of synapses in embryonic rat hippocampal neurons *in vitro*. J Neurochem, 95, 13731386.

Katanaev VL, Solis GP, Hausmann G, Buestorf S, Katanayeva N, Schrock Y, Stuermer CA, and Basler K (2008) Reggie-1/flotillin-2 promotes secretion of the long-range signalling forms of Wingless and Hedgehog in *Drosophila*. EMBO J, 27, 509-521.

Khosravani H, Zhang Y, Tsutsui S, Hameed S, Altier C, Hamid J, Chen L, Villemaire M, Ali Z, Jirik FR, and Zamponi GW (2008) Prion protein attenuates excitotoxicity by inhibiting NMDA receptors. J Cell Biol, 181, 551-565.

Kleene R, Loers G, Langer J, Frobert Y, Buck F, and Schachner M (2007) Prion protein regulates glutamate-dependent lactate transport of astrocytes. J Neurosci, 27, 12331-12340.

Kocisko DA, Vaillant A, Lee KS, Arnold KM, Bertholet N, Race RE, Olsen EA, Juteau JM, and Caughey B (2006) Potent antiscrapie activities of degenerate phosphorothioate oligonucleotides. Antimicrob Agents Chemother, 50, 10341044.

Kong Q, Surewicz WK, Petersen RB, Zou W, Chen SG, Gambetti P, Parchi P, Capellari S, Goldfarb L, Montagna P, Lugaresi E, Piccardo P, and Ghetti B (2003). Inherited Prion Diseases. In Stanley B Prusiner (Ed.), Prion Biology and Diseases, . John Inglis, pp. 673-775.

Krieglstein K and Unsicker K (1996) Proteins from chromaffin granules promote survival of dorsal root ganglionic neurons: comparison with neurotrophins. Brain Res Dev Brain Res, 93, 10-17.

Kristensson K, Feuerstein B, Taraboulos A, Hyun WC, Prusiner SB, and DeArmond SJ (1993) Scrapie prions alter receptor-mediated calcium responses in cultured cells. Neurology, 43, 2335-2341.

Kurisaki A, Inoue I, Kurisaki K, Yamakawa N, Tsuchida K, and Sugino H (2008) Activin induces long-lasting N-methyl-D-aspartate receptor activation via scaffolding PDZ protein activin receptor interacting protein 1. Neuroscience, 151, 1225-1235.

Kuwahara C, Takeuchi AM, Nishimura T, Haraguchi K, Kubosaki A, Matsumoto Y, Saeki K, Matsumoto Y, Yokoyama T, Itohara S, and Onodera T (1999) Prions prevent neuronal cell-line death. Nature, 400, 225-226.

Laffont-Proust I, Faucheux BA, Hassig R, Sazdovitch V, Simon S, Grassi J, Hauw JJ, Moya KL, and Haik S (2005) The N-terminal cleavage of cellular prion protein in the human brain. FEBS Lett, 579, 6333-6337.

Landolt HP, Glatzel M, Blattler T, Achermann P, Roth C, Mathis J, Weis J, Tobler I, Aguzzi A, and Bassetti CL (2006) Sleep-wake disturbances in sporadic Creutzfeldt-Jakob disease. Neurology, 66, 1418-1424.

Lauren J, Gimbel DA, Nygaard HB, Gilbert JW, and Strittmatter SM (2009) Cellular prion protein mediates impairment of synaptic plasticity by amyloid-beta oligomers. Nature, 457, 1128-1132.

Lee FS, Rajagopal R, Kim AH, Chang PC, and Chao MV (2002) Activation of Trk neurotrophin receptor signaling by pituitary adenylate cyclase-activating polypeptides. J Biol Chem, 277, 9096-9102.

Lee KJ, Panzera A, Rogawski D, Greene LE, and Eisenberg E (2007a) Cellular prion protein (PrPC) protects neuronal cells from the effect of huntingtin aggregation. J Cell Sci, 120, 2663-2671.

Lee KS, Magalhaes AC, Zanata SM, Brentani RR, Martins VR, and Prado MA (2001) Internalization of mammalian fluorescent cellular prion protein and N-terminal deletion mutants in living cells. J Neurochem, 79, 79-87.

Lee KS, Raymond LD, Schoen B, Raymond GJ, Kett L, Moore RA, Johnson LM, Taubner L, Speare JO, Onwubiko HA, Baron GS, Caughey WS, and Caughey B (2007b) Hemin interactions and alterations of the subcellular localization of prion protein. J Biol Chem, 282, 36525-36533.

Lefebvre-Roque M, Kremmer E, Gilch S, Zou WQ, Feraudet C, Gilles CM, Sales N, Grassi J, Gambetti P, Baron T, Schatzl H, and Lasmezas CI (2007) Toxic effects of intracerebral PrP antibody administration during the course of BSE infection in mice. Prion, 1, 198-206.

Levi-Montalcini R (1987) The nerve growth factor 35 years later. Science, 237, 1154-1162.

Levi-Montalcini R and Calissano P (2006) The scientific challenge of the 21st century: from a reductionist to a holistic approach via systems biology. BMC Neurosci, 7 Suppl 1, S1.

Levi-Montalcini R, Skaper SD, Dal TR, Petrelli L, and Leon A (1996) Nerve growth factor: from neurotrophin to neurokine. Trends Neurosci, 19, 514-520.

Levine JM (1992) Historical perspective: the neurotrophic theory of skin ulceration. J Am Geriatr Soc, 40, 1281-1283.

Li A, Christensen HM, Stewart LR, Roth KA, Chiesa R, and Harris DA (2007) Neonatal lethality in transgenic mice expressing prion protein with a deletion of residues 105-125. EMBO J, 26, 548-558.

Liesi P, Narvanen A, Soos J, Sariola H, and Snounou G (1989) Identification of a neurite outgrowth-promoting domain of laminin using synthetic peptides. FEBS Lett, 244, 141-148.

Lima FR, Arantes CP, Muras AG, Nomizo R, Brentani RR, and Martins VR (2007) Cellular prion protein expression in astrocytes modulates neuronal survival and differentiation. J Neurochem. 103, 2164-2176.

Linden R (1994) The survival of developing neurons: a review of afferent control. Neuroscience, 58, 671-682.

Linden R, Martins RA, and Silveira MS (2005) Control of programmed cell death by neurotransmitters and neuropeptides in the developing mammalian retina. Prog Retin Eye Res, 24, 457-491.

Linden R, Martins VR, Prado MA, Cammarota M, Izquierdo I, and Brentani RR (2008) Physiology of the prion protein. Physiol Rev, 88, 673-728.

Lopes MH, Hajj GN, Muras AG, Mancini GL, Castro RM, Ribeiro KC, Brentani RR, Linden R, and Martins VR (2005) Interaction of cellular prion and stress-inducible protein 1 promotes neuritogenesis and neuroprotection by distinct signaling pathways. J Neurosci, 25, 11330-11339.

Lopez GF, Zahn R, Riek R, and Wuthrich K (2000) NMR structure of the bovine prion protein. Proc Natl Acad Sci U S A, 97, 8334-8339.

Luckenbill-Edds L (1997) Laminin and the mechanism of neuronal outgrowth. Brain Res Brain Res Rev, 23, 1-27.

Lysek DA, Schorn C, Nivon LG, Esteve-Moya V, Christen B, Calzolai L, von SC, Fiorito F, Herrmann T, Guntert P, and Wuthrich K (2005) Prion protein NMR structures of cats, dogs, pigs, and sheep. Proc Natl Acad Sci U S A, 102, 640-645.

Magalhaes AC, Silva JA, Lee KS, Martins VR, Prado VF, Ferguson SS, Gomez MV, Brentani RR, and Prado MA (2002) Endocytic intermediates involved with the intracellular trafficking of a fluorescent cellular prion protein. J Biol Chem, 277, 33311-33318.

Maglio LE, Perez MF, Martins VR, Brentani RR, and Ramirez OA (2004) Hippocampal synaptic plasticity in mice devoid of cellular prion protein. Brain Res Mol Brain Res, 131, 58-64.

Malaga-Trillo E, Solis GP, Schrock Y, Geiss C, Luncz L, Thomanetz V, and Stuermer CA (2009) Regulation of embryonic cell adhesion by the prion protein. PLoS Biol, 7, e55.

Mallucci GR, Ratte S, Asante EA, Linehan J, Gowland I, Jefferys JG, and Collinge J (2002) Post-natal knockout of prion protein alters hippocampal CA1 properties, but does not result in neurodegeneration. EMBO J, 21, 202-210.

Mallucci G, Dickinson A, Linehan J, Klohn PC, Brandner S, and Collinge J (2003) Depleting neuronal PrP in prion infection prevents disease and reverses spongiosis. Science, 302, 871-874.

Mallucci GR, White MD, Farmer M, Dickinson A, Khatun H, Powell AD, Brandner S, Jefferys JG, and Collinge J (2007) Targeting cellular prion protein reverses early cognitive deficits and neurophysiological dysfunction in prion-infected mice. Neuron, 53, 325-335.

Manson JC, Clarke AR, Hooper ML, Aitchison L, McConnell I, and Hope J (1994) 129/Ola mice carrying a null mutation in PrP that abolishes mRNA production are developmentally normal. Mol Neurobiol, 8, 121-127.

Marchi S, Rimessi A, Giorgi C, Baldini C, Ferroni L, Rizzuto R, and Pinton P (2008) Akt kinase reducing endoplasmic reticulum Ca2+ release protects cells from Ca2+-dependent apoptotic stimuli. Biochem Biophys Res Commun, 375, 501-505.

Martins AH, Resende RR, Majumder P, Faria M, Casarini DE, Tarnok A, Colli W, Pesquero JB, and Ulrich H (2005) Neuronal differentiation of P19 embryonal carcinoma cells modulates kinin B2 receptor gene expression and function. J Biol Chem, 280, 19576-19586.

Martins RA and Pearson RA (2008) Control of cell proliferation by neurotransmitters in the developing vertebrate retina. Brain Res, 1192, 37-60.

Martins VR, Graner E, Garcia-Abreu J, de Souza SJ, Mercadante AF, Veiga SS, Zanata SM, Neto VM, and Brentani RR (1997) Complementary hydropathy identifies a cellular prion protein receptor. Nat Med, 3, 1376-1382.

Martins VR, Linden R, Prado MA, Walz R, Sakamoto AC, Izquierdo I, and Brentani RR (2002) Cellular prion protein: on the road for functions. FEBS Lett, 512, 25-28.

Masters CL, Gajdusek DC, and Gibbs CJ, Jr. (1981) Creutzfeldt-Jakob disease virus isolations from the Gerstmann-Straussler syndrome with an analysis of the various forms of amyloid plaque deposition in the virus-induced spongiform encephalopathies. Brain, 104, 559-588.

Mattei V, Barenco MG, Tasciotti V, Garofalo T, Longo A, Boller K, Lower J, Misasi R, Montrasio F, and Sorice M (2009) Paracrine diffusion of PrP(C) and propagation of prion infectivity by plasma membrane-derived microvesicles. PLoS One, 4, e5057.

McLennan NF, Brennan PM, McNeill A, Davies I, Fotheringham A, Rennison KA, Ritchie D, Brannan F, Head MW, Ironside JW, Williams A, and Bell JE (2004) Prion protein accumulation and neuroprotection in hypoxic brain damage. Am J Pathol, 165, 227-235.

Meldrum BS (2000) Glutamate as a neurotransmitter in the brain: review of physiology and pathology. J Nutr, 130, 1007S-1015S.

Middleton G, Hamanoue M, Enokido Y, Wyatt S, Pennica D, Jaffray E, Hay RT, and Davies AM (2000) Cytokine-induced nuclear factor kappa B activation promotes the survival of developing neurons. J Cell Biol, 148, 325-332.

Mitsios N, Saka M, Krupinski J, Pennucci R, Sanfeliu C, Miguel TM, Gaffney J, Kumar P, Kumar S, Sullivan M, and Slevin M (2007) Cellular prion protein is increased in the

plasma and peri-infarcted brain tissue after acute stroke. J Neurosci Res, 85, 602-611.
Moreno DH (1999) Molecular and functional diversity of voltage-gated calcium channels. Ann N Y Acad Sci, 868, 102-117.
Morillas M, Swietnicki W, Gambetti P, and Surewicz WK (1999) Membrane environment alters the conformational structure of the recombinant human prion protein. J Biol Chem, 274, 36859-36865.
Mouillet-Richard S, Ermonval M, Chebassier C, Laplanche JL, Lehmann S, Launay JM, and Kellermann O (2000) Signal transduction through prion protein. Science, 289, 1925-1928.
Mouillet-Richard S, Pietri M, Schneider B, Vidal C, Mutel V, Launay JM, and Kellermann O (2005) Modulation of serotonergic receptor signaling and crosstalk by prion protein. J Biol Chem, 280, 4592-4601.
Mouillet-Richard S, Schneider B, Pradines E, Pietri M, Ermonval M, Grassi J, Richards JG, Mutel V, Launay JM, and Kellermann O (2007) Cellular prion protein signaling in serotonergic neuronal cells. Ann N Y Acad Sci, 1096, 106-119.
Muramoto T, DeArmond SJ, Scott M, Telling GC, Cohen FE, and Prusiner SB (1997) Heritable disorder resembling neuronal storage disease in mice expressing prion protein with deletion of an alpha-helix. Nat Med, 3, 750-755.
Murrin LC and Talbot JN (2007) RanBPM, a scaffolding protein in the immune and nervous systems. J Neuroimmune Pharmacol, 2, 290-295.
Nagappan G and Lu B (2005) Activity-dependent modulation of the BDNF receptor TrkB: mechanisms and implications. Trends Neurosci, 28, 464-471.
Nakaoka Y, Nishida K, Fujio Y, Izumi M, Terai K, Oshima Y, Sugiyama S, Matsuda S, Koyasu S, Yamauchi-Takihara K, Hirano T, Kawase I, and Hirota H (2003) Activation of gp130 transduces hypertrophic signal through interaction of scaffolding/docking protein Gab1 with tyrosine phosphatase SHP2 in cardiomyocytes. Circ Res, 93, 221-229.
Naslavsky N, Stein R, Yanai A, Friedlander G, and Taraboulos A (1997) Characterization of detergent-insoluble complexes containing the cellular prion protein and its scrapie isoform. J Biol Chem, 272, 6324-6331.
Nazor KE, Seward T, and Telling GC (2007) Motor behavioral and neuropathological deficits in mice deficient for normal prion protein expression. Biochim Biophys Acta, 1772, 645-653.
Nelson J, McFerran NV, Pivato G, Chambers E, Doherty C, Steele D, and Timson DJ (2008) The 67 kDa laminin receptor: structure, function and role in disease. Biosci Rep, 28, 33-48.
Nichols BJ, Kenworthy AK, Polishchuk RS, Lodge R, Roberts TH, Hirschberg K, Phair RD, and Lippincott-Schwartz J (2001) Rapid cycling of lipid raft markers between the cell surface and Golgi complex. J Cell Biol, 153, 529-541.
Nishimura T, Sakudo A, Nakamura I, Lee DC, Taniuchi Y, Saeki K, Matsumoto Y, Ogawa M, Sakaguchi S, Itohara S, and Onodera T (2004) Cellular prion protein regulates intracellular hydrogen peroxide level and prevents copper-induced apoptosis. Biochem Biophys Res Commun, 323, 218-222.
Nykjaer A, Petersen CM, Moller B, Jensen PH, Moestrup SK, Holtet TL, Etzerodt M, Thogersen HC, Munch M, Andreasen PA, and . (1992) Purified alpha 2macroglobulin receptor/LDL receptor-related protein binds urokinase. plasminogen activator inhibitor type-1 complex. Evidence that the alpha 2-macroglobulin receptor mediates cellular degradation of urokinase receptor-bound complexes. J Biol Chem, 267, 14543-14546.
Oppenheim RW (1989) The neurotrophic theory and naturally occurring motoneuron death. Trends Neurosci, 12, 252-255.
Pantera B, Bini C, Cirri P, Paoli P, Camici G, Manao G, and Caselli A (2009) PrPc activation induces neurite outgrowth and differentiation in PC12 cells: role for caveolin-1 in the signal transduction pathway. J Neurochem, 110, 194-207.
Parizek P, Roeckl C, Weber J, Flechsig E, Aguzzi A, and Raeber AJ (2001) Similar turnover and shedding of the cellular prion protein in primary lymphoid and neuronal cells. J Biol Chem, 276, 44627-44632.
Parkin ET, Watt NT, Hussain I, Eckman EA, Eckman CB, Manson JC, Baybutt HN, Turner AJ, and Hooper NM (2007) Cellular prion protein regulates betasecretase cleavage of the Alzheimer's amyloid precursor protein. Proc Natl Acad Sci U S A, 104, 11062-11067.
Parkin ET, Watt NT, Turner AJ, and Hooper NM (2004) Dual mechanisms for shedding of the cellular prion protein. J Biol Chem, 279, 11170-11178.
Parkyn CJ, Vermeulen EG, Mootoosamy RC, Sunyach C, Jacobsen C, Oxvig C, Moestrup S, Liu Q, Bu G, Jen A, and Morris RJ (2008) LRP1 controls biosynthetic and endocytic trafficking of neuronal prion protein. J Cell Sci, 121, 773-783.
Patel AJ and Honore E (2001) Properties and modulation of mammalian 2P domain K+ channels. Trends Neurosci, 24, 339-346.
Pauly PC and Harris DA (1998) Copper stimulates endocytosis of the prion protein. J Biol Chem, 273, 33107-33110.
Pawson T and Scott JD (1997) Signaling through scaffold, anchoring, and adaptor proteins. Science, 278, 2075-2080.
Perera WS and Hooper NM (2001) Ablation of the metal ion-induced endocytosis of the prion protein by disease-associated mutation of the octarepeat region. Curr Biol, 11, 519-523.
Perini F, Frangione B, and Prelli F (1996) Prion protein released by platelets. Lancet, 347, 1635-1636.
Ping A, Chun ZX, and Xue XY (2005) Bradykinin preconditioning induces protective effects against focal cerebral ischemia in rats. Brain Res, 1059, 105-112.
Pires Neto MA, Braga-de-Souza S, and Lent R (1999) Extracellular matrix molecules play diverse roles in the growth and guidance of central nervous system axons. Braz J Med Biol Res, 32, 633-638.
Pons S and Marti E (2000) Sonic hedgehog synergizes with the extracellular matrix prote*in vitro*nectin to induce spinal motor neuron differentiation. Development, 127, 333-342.
Pons S, Trejo JL, Martinez-Morales JR, and Marti E (2001) Vitronectin regulates Sonic hedgehog activity during cerebellum development through CREB phosphorylation. Development, 128, 1481-1492.
Pradines E, Loubet D, Schneider B, Launay JM, Kellermann O, and Mouillet-Richard S (2008) CREB-dependent gene

regulation by prion protein: Impact on MMP-9 and beta-dystroglycan. Cell Signal, 20, 2050-2058.

Prado MA, ves-Silva J, Magalhaes AC, Prado VF, Linden R, Martins VR, and Brentani RR (2004) PrPc on the road: trafficking of the cellular prion protein. J Neurochem, 88, 769-781.

Prestori F, Rossi P, Bearzatto B, Laine J, Necchi D, Diwakar S, Schiffmann SN, Axelrad H, and D'Angelo E (2008) Altered neuron excitability and synaptic plasticity in the cerebellar granular layer of juvenile prion protein knock-out mice with impaired motor control. J Neurosci, 28, 7091-7103.

Priola SA and Caughey B (1994) Inhibition of scrapie-associated PrP accumulation. Probing the role of glycosaminoglycans in amyloidogenesis. Mol Neurobiol, 8, 113-120.

Provini F, Vetrugno R, Pierangeli G, Cortelli P, Rizzo G, Filla A, Strisciuglio C, Gallassi R, and Montagna P (2008) Sleep and temperature rhythms in two sisters with P102L Gerstmann-Straussler-Scheinker (GSS) disease. Sleep Med. 10, 374-377.

Radovanovic I, Braun N, Giger OT, Mertz K, Miele G, Prinz M, Navarro B, and Aguzzi A (2005) Truncated prion protein and Doppel are myelinotoxic in the absence of oligodendrocytic PrPC. J Neurosci, 25, 4879-4888.

Rambold AS, Muller V, Ron U, Ben-Tal N, Winklhofer KF, and Tatzelt J (2008) Stress-protective signalling of prion protein is corrupted by scrapie prions. EMBO J, 27, 1974-1984.

Rangel A, Burgaya F, Gavin R, Soriano E, Aguzzi A, and Del Rio JA (2007) Enhanced susceptibility of Prnp-deficient mice to kainate-induced seizures, neuronal apoptosis, and death: Role of AMPA/kainate receptors. J Neurosci Res, 85, 2741-2755.

Rauch U (2004) Extracellular matrix components associated with remodeling processes in brain. Cell Mol Life Sci, 61, 2031-2045.

Rieger R, Edenhofer F, Lasmezas CI, and Weiss S (1997) The human 37-kDa laminin receptor precursor interacts with the prion protein in eukaryotic cells. Nat Med, 3, 1383-1388.

Riek R, Hornemann S, Wider G, Glockshuber R, and Wuthrich K (1997) NMR characterization of the full-length recombinant murine prion protein, mPrP(23231). FEBS Lett, 413, 282-288.

Robertson C, Booth SA, Beniac DR, Coulthart MB, Booth TF, and McNicol A (2006) Cellular prion protein is released on exosomes from activated platelets. Blood, 107, 3907-3911.

Rodriguez MM, Peoc'h K, Haik S, Bouchet C, Vernengo L, Manana G, Salamano R, Carrasco L, Lenne M, Beaudry P, Launay JM, and Laplanche JL (2005) A novel mutation (G114V) in the prion protein gene in a family with inherited prion disease. Neurology, 64, 1455-1457.

Rudd PM, Endo T, Colominas C, Groth D, Wheeler SF, Harvey DJ, Wormald MR, Serban H, Prusiner SB, Kobata A, and Dwek RA (1999) Glycosylation differences between the normal and pathogenic prion protein isoforms. Proc Natl Acad Sci U S A, 96, 13044-13049.

Runeberg-Roos P and Saarma M (2007) Neurotrophic factor receptor RET: structure, cell biology, and inherited diseases. Ann Med, 39, 572-580.

Sadowski MJ, Pankiewicz J, Prelli F, Scholtzova H, Spinner DS, Kascsak RB, Kascsak RJ, and Wisniewski T (2009) Anti-PrP Mab 6D11 suppresses PrP(Sc) replication in prion infected myeloid precursor line FDC-P1/22L and in the lymphoreticular system in vivo. Neurobiol Dis, 34, 267-278.

Sadri-Vakili G and Cha JH (2006) Histone deacetylase inhibitors: a novel therapeutic approach to Huntington's disease (complex mechanism of neuronal death). Curr Alzheimer Res, 3, 403-408.

Samaia HB and Brentani RR (1998) Can loss-of-function prion-related diseases exist? Mol Psychiatry, 3, 196-197.

Sanford SD, Gatlin JC, Hokfelt T, and Pfenninger KH (2008) Growth cone responses to growth and chemotropic factors. Eur J Neurosci, 28, 268-278.

Santuccione A, Sytnyk V, Leshchyns'ka I, and Schachner M (2005) Prion protein recruits its neuronal receptor NCAM to lipid rafts to activate p59fyn and to enhance neurite outgrowth. J Cell Biol, 169, 341-354.

Sariola H and Saarma M (2003) Novel functions and signalling pathways for GDNF. J Cell Sci, 116, 3855-3862.

Schmitt-Ulms G, Legname G, Baldwin MA, Ball HL, Bradon N, Bosque PJ, Crossin KL, Edelman GM, DeArmond SJ, Cohen FE, and Prusiner SB (2001) Binding of neural cell adhesion molecules (N-CAMs) to the cellular prion protein. J Mol Biol, 314, 1209-1225.

Schonberger O, Horonchik L, Gabizon R, Papy-Garcia D, Barritault D, and Taraboulos A (2003) Novel heparan mimetics potently inhibit the scrapie prion protein and its endocytosis. Biochem Biophys Res Commun, 312, 473-479.

Schrock Y, Solis GP, and Stuermer CA (2009) Regulation of focal adhesion formation and filopodia extension by the cellular prion protein (PrPC). FEBS Lett, 583, 389-393.

Sharma HS (2007) A select combination of neurotrophins enhances neuroprotection and functional recovery following spinal cord injury. Ann N Y Acad Sci, 1122, 95-111.

Shmerling D, Hegyi I, Fischer M, Blattler T, Brandner S, Gotz J, Rulicke T, Flechsig E, Cozzio A, von MC, Hangartner C, Aguzzi A, and Weissmann C (1998) Expression of amino-terminally truncated PrP in the mouse leading to ataxia and specific cerebellar lesions. Cell, 93, 203-214.

Shyng SL, Heuser JE, and Harris DA (1994) A glycolipid-anchored prion protein is endocytosed via clathrin-coated pits. J Cell Biol, 125, 1239-1250.

Shyu WC, Lin SZ, Chiang MF, Ding DC, Li KW, Chen SF, Yang HI, and Li H (2005) Overexpression of PrPC by adenovirus-mediated gene targeting reduces ischemic injury in a stroke rat model. J Neurosci, 25, 8967-8977.

Silva JL, Lima LM, Foguel D, and Cordeiro Y (2008) Intriguing nucleic-acidbinding features of mammalian prion protein. Trends Biochem Sci, 33, 132-140.

Solforosi L, Criado JR, McGavern DB, Wirz S, Sanchez-Alavez M, Sugama S, DeGiorgio LA, Volpe BT, Wiseman E, Abalos G, Masliah E, Gilden D, Oldstone MB, Conti B, and Williamson RA (2004) Cross-linking cellular prion protein triggers neuronal apoptosis in vivo. Science, 303, 1514-1516.

Spudich A, Frigg R, Kilic E, Kilic U, Oesch B, Raeber A, Bassetti CL, and Hermann DM (2005) Aggravation of ischemic brain injury by prion protein deficiency: role of ERK-1/-2 and STAT-1. Neurobiol Dis, 20, 442-449.

Stuermer CA, Langhorst MF, Wiechers MF, Legler DF, Von Hanwehr SH, Guse AH, and Plattner H (2004) PrPc capping in T cells promotes its association with the lipid raft proteins reggie-1 and reggie-2 and leads to signal transduction. FASEB J, 18, 1731-1733.

Stuermer CA and Plattner H (2005) The 'lipid raft' microdomain proteins reggie-1 and reggie-2 (flotillins) are scaffolds for protein interaction and signalling. Biochem Soc Symp, 109-118.

Sunyach C, Cisse MA, da Costa CA, Vincent B, and Checler F (2007) The C-terminal products of cellular prion protein processing, C1 and C2, exert distinct influence on p53-dependent staurosporine-induced caspase-3 activation. J Biol Chem, 282, 1956-1963.

Sunyach C, Jen A, Deng J, Fitzgerald KT, Frobert Y, Grassi J, McCaffrey MW, and Morris R (2003) The mechanism of internalization of glycosylphosphatidylinositol-anchored prion protein. EMBO J, 22, 3591-3601.

Tagliavini F, Prelli F, Porro M, Salmona M, Bugiani O, and Frangione B (1992) A soluble form of prion protein in human cerebrospinal fluid: implications for prion-related encephalopathies. Biochem Biophys Res Commun, 184, 1398-1404.

Tayebi M and Hawke S (2006) Antibody-mediated neuronal apoptosis: therapeutic implications for prion diseases. Immunol Lett, 105, 123-126.

Taylor DR and Hooper NM (2007) The low-density lipoprotein receptor-related protein 1 (LRP1) mediates the endocytosis of the cellular prion protein. Biochem J, 402, 17-23.

Telling GC, Haga T, Torchia M, Tremblay P, DeArmond SJ, and Prusiner SB (1996) Interactions between wild-type and mutant prion proteins modulate neurodegeneration in transgenic mice. Genes Dev, 10, 1736-1750.

Theodosis DT, Piet R, Poulain DA, and Oliet SH (2004) Neuronal, glial and synaptic remodeling in the adult hypothalamus: functional consequences and role of cell surface and extracellular matrix adhesion molecules. Neurochem Int, 45, 491-501.

Theodosis DT, Poulain DA, and Oliet SH (2008) Activity-dependent structural and functional plasticity of astrocyte-neuron interactions. Physiol Rev, 88, 9831008.

Thiel G and Cibelli G (2002) Regulation of life and death by the zinc finger transcription factor Egr-1. J Cell Physiol, 193, 287-292.

Thoenen H, Hughes RA, and Sendtner M (1993) Trophic support of motoneurons: physiological, pathophysiological, and therapeutic implications. Exp Neurol, 124, 47-55.

Tobler I, Deboer T, and Fischer M (1997) Sleep and sleep regulation in normal and prion protein-deficient mice. J Neurosci, 17, 1869-1879.

Tobler I, Gaus SE, Deboer T, Achermann P, Fischer M, Rulicke T, Moser M, Oesch B, McBride PA, and Manson JC (1996) Altered circadian activity rhythms and sleep in mice devoid of prion protein. Nature, 380, 639-642.

Tzu J and Marinkovich MP (2008) Bridging structure with function: structural, regulatory, and developmental role of laminins. Int J Biochem Cell Biol, 40, 199-214.

Vana K and Weiss S (2006) A trans-dominant negative 37kDa/67kDa laminin receptor mutant impairs PrP(Sc) propagation in scrapie-infected neuronal cells. J Mol Biol, 358, 57-66.

Vella LJ, Greenwood DL, Cappai R, Scheerlinck JP, and Hill AF (2008) Enrichment of prion protein in exosomes derived from ovine cerebral spinal fluid. Vet Immunol Immunopathol, 124, 385-393.

Vey M, Pilkuhn S, Wille H, Nixon R, DeArmond SJ, Smart EJ, Anderson RG, Taraboulos A, and Prusiner SB (1996) Subcellular colocalization of the cellular and scrapie prion proteins in caveolae-like membranous domains. Proc Natl Acad Sci U S A, 93, 14945-14949.

Viard P, Butcher AJ, Halet G, Davies A, Nurnberg B, Heblich F, and Dolphin AC (2004) PI3K promotes voltage-dependent calcium channel trafficking to the plasma membrane. Nat Neurosci, 7, 939-946.

Vincent B, Cisse MA, Sunyach C, Guillot-Sestier MV, and Checler F (2008) Regulation of betaAPP and PrPc cleavage by alpha-secretase: mechanistic and therapeutic perspectives. Curr Alzheimer Res, 5, 202-211.

Vincent B, Paitel E, Frobert Y, Lehmann S, Grassi J, and Checler F (2000) Phorbol ester-regulated cleavage of normal prion protein in HEK293 human cells and murine neurons. J Biol Chem, 275, 35612-35616.

Vincent B, Paitel E, Saftig P, Frobert Y, Hartmann D, De SB, Grassi J, Lopez-Perez E, and Checler F (2001) The disintegrins ADAM10 and TACE contribute to the constitutive and phorbol ester-regulated normal cleavage of the cellular prion protein. J Biol Chem, 276, 37743-37746.

Walz R, Amaral OB, Rockenbach IC, Roesler R, Izquierdo I, Cavalheiro EA, Martins VR, and Brentani RR (1999) Increased sensitivity to seizures in mice lacking cellular prion protein. Epilepsia, 40, 1679-1682.

Warner RG, Hundt C, Weiss S, and Turnbull JE (2002) Identification of the heparan sulfate binding sites in the cellular prion protein. J Biol Chem, 277, 18421-18430.

Weise J, Crome O, Sandau R, Schulz-Schaeffer W, Bahr M, and Zerr I (2004) Upregulation of cellular prion protein (PrPc) after focal cerebral ischemia and influence of lesion severity. Neurosci Lett, 372, 146-150.

Weise J, Doeppner TR, Muller T, Wrede A, Schulz-Schaeffer W, Zerr I, Witte OW, and Bahr M (2008) Overexpression of cellular prion protein alters postischemic Erk1/2 phosphorylation but not Akt phosphorylation and protects against focal cerebral ischemia. Restor Neurol Neurosci, 26, 57-64.

Weise J, Sandau R, Schwarting S, Crome O, Wrede A, Schulz-Schaeffer W, Zerr I, and Bahr M (2006) Deletion of cellular prion protein results in reduced Akt activation, enhanced postischemic caspase-3 activation, and exacerbation of ischemic brain injury. Stroke, 37, 1296-1300.

Whatley SA, Powell JF, Politopoulou G, Campbell IC, Brammer MJ, and Percy NS (1995) Regulation of intracellular free calcium levels by the cellular prion protein. Neuroreport, 6, 2333-2337.

White AR, Enever P, Tayebi M, Mushens R, Linehan J, Brandner S, Anstee D, Collinge J, and Hawke S (2003) Monoclonal antibodies inhibit prion replication and delay the development of prion disease. Nature, 422, 80-83.

White MD, Farmer M, Mirabile I, Brandner S, Collinge J, and Mallucci GR (2008) Single treatment with RNAi against prion protein rescues early neuronal dysfunction and prolongs survival in mice with prion disease. Proc Natl Acad Sci U S A, 105,

Wiese S, Jablonka S, Holtmann B, Orel N, Rajagopal R, Chao MV, and Sendtner M (2007) Adenosine receptor A2A-R contributes to motoneuron survival by transactivating the tyrosine kinase receptor TrkB. Proc Natl Acad Sci U S A, 104, 17210-17215.

Williams GR (2008) Neurodevelopmental and neurophysiological actions of thyroid hormone. J Neuroendocrinol, 20, 784-794.

Wong K, Qiu Y, Hyun W, Nixon R, VanCleff J, Sanchez-Salazar J, Prusiner SB, and DeArmond SJ (1996) Decreased receptor-mediated calcium response in prion-infected cells correlates with decreased membrane fluidity and IP3 release. Neurology, 47, 741-750.

Yan-Feng W, Gang L, and Yan-Ting G (2008) Bradykinin preconditioning induces protective effects on the spinal cord ischemic injury of rats. Neurosci Lett, 433, 114-118.

Yano H and Chao MV (2000) Neurotrophin receptor structure and interactions. Pharm Acta Helv, 74, 253-260.

Ye J, Han J, Shi Q, Zhang BY, Wang GR, Tian C, Gao C, Chen JM, Li CJ, Lui Z, Li XZ, Zhang LZ, and Dong XP (2008) A genetic human prion disease with G114V mutation in PRNP and epidemiological studies in a Chinese family: a case series. J Med Case Reports, 2, 331.

Zahn R, Liu A, Luhrs T, Riek R, von SC, Lopez GF, Billeter M, Calzolai L, Wider G, and Wuthrich K (2000) NMR solution structure of the human prion protein. Proc Natl Acad Sci U S A, 97, 145-150.

Zanata SM, Lopes MH, Mercadante AF, Hajj GN, Chiarini LB, Nomizo R, Freitas AR, Cabral AL, Lee KS, Juliano MA, de Oliveira E, Jachieri SG, Burlingame A, Huang L, Linden R, Brentani RR, and Martins VR (2002) Stress-inducible protein 1 is a cell surface ligand for cellular prion that triggers neuroprotection. EMBO J, 21, 3307-3316.

Zhang J, Zhang D, McQuade JS, Behbehani M, Tsien JZ, and Xu M (2002) c-fos regulates neuronal excitability and survival. Nat Genet, 30, 416-420.

Zuber C, Knackmuss S, Rey C, Reusch U, Rottgen P, Frohlich T, Arnold GJ, Pace C, Mitteregger G, Kretzschmar HA, Little M, and Weiss S (2008) Single chain Fv antibodies directed against the 37 kDa/67 kDa laminin receptor as therapeutic tools in prion diseases. Mol Immunol, 45, 144-151.

Chapter 3

Autophagy, Prion Infection and their Mutual Interactions

Andreas Heiseke[1], Yasmine Aguib[1] and Hermann M. Schatzl

Abstract
Prion diseases are infectious and fatal neurodegenerative disorders of man and animals which are characterized by spongiform degeneration in the central nervous system. Prion propagation involves the endocytic pathway and endosomal and lysosomal compartments are implicated in trafficking and re-cycling as well as final degradation of prions. Shifting the equilibrium between propagation and lysosomal clearance to the latter impairs cellular prion load. This and earlier findings of autophagic vacuoles in correlation to prion infections both in *in vitro* and *in vivo* studies prompted us and others to analyze the role of autophagy in prion infection. Autophagy is a fundamental cellular bulk degradation process for e.g. organelles or cytoplasmic proteins which has many implications for physiology and patho-physiology of cells and whole organisms. In various neurodegenerative disease models mainly protective functions of autophagy were recently described. In this review, we focus on recent findings which correlate autophagy and its manipulations with prion infection scenarios, and discuss perspectives and future directions. The findings summarized here add to the knowledge of the role of autophagy in neurodegeneration and provide interesting new insight into how non-cytosolic aggregated proteins might be subjected to autophagic clearance.

Introduction
Prion diseases are infectious neurodegenerative disorders that can affect humans and animals. Examples for animal prion diseases are scrapie in sheep and goat, bovine spongiform encephalopathy (BSE) in cattle, transmissible mink encephalopathy (TME) in mink, and chronic wasting disease (CWD) in elk and deer. Prion diseases in humans are Creutzfeldt-Jakob disease (CJD), Gerstmann-Sträussler-Scheinker (GSS) syndrome, fatal familiar insomnia (FFI), kuru, and the new variants of CJD (vCJD and secondary vCJD). The disease is characterized by a rapidly progressing course that leads inevitably to death, usually within a few months. Typically, this is preceded by a long incubation time entirely free of symptoms, lasting for years to many decades in humans. Severe loss of neurons is a key characteristic for all prion diseases, accompanied by strong astrogliosis and mild microglia activation. This results in a progressive spongiform degeneration of the central nervous system (CNS) which manifests itself in ataxia, behavioral changes and, in humans, a highly progressive loss of intellectual abilities (Aguzzi and Polymenidou, 2004; Collinge, 2005; Prusiner, 1998; Weissmann, 2004).

According to the protein-only hypothesis these diseases are caused by prions, *proteinaceous infectious* particles devoid of encoding nucleic acid (Prusiner, 1982). They consist mainly, if not solely, of an abnormally folded isoform (PrP^{Sc}) of the normal, host-encoded prion protein PrP^{c} (Aguzzi and Polymenidou, 2004; Cohen et al., 1994; Prusiner, 1998). Prions have self-propagating capacities in that they are able to catalyze a profound conformational switch from PrP^{c} into an aggregated structure resulting finally in the accumulation of misfolded and aggregated PrP^{Sc} in the brain. Hence, prion diseases share profound similarities with other protein misfolding and neurodegenerative diseases like Alzheimer's, Huntington's and Parkinson's disease (Aguzzi and Haass, 2003). Yet, prions are unique as they are not only able to replicate their conformation but are also naturally and experimentally transmissible within and to some extent between species.

In prion-infected cultured neuronal cells (Schatzl et al., 1997) and also in brain biopsy materials of prion-infected patients (Liberski et al., 2008; Sikorska et al., 2004) the appearance of multi-vesicular bodies and autophagic vacuoles has been reported. This prompted us and others to investigate in detail whether there is a link between autophagy and prion infection, in particular as the impact of autophagy in various other neurodegenerative diseases has been recently described (Bursch and Ellinger, 2005; Rubinsztein, 2006).

In this review we focus on the cellular mechanism of autophagy, its regulation, some well-established methods to monitor autophagy, its implications in disease with emphasis on neurodegenerative disorders, and finally, its putative role in prion infection scenarios.

Introducing autophagy
Degradation of organelles or cytoplasmic proteins can be mediated by an intracellular bulk degradation process called macroautophagy (referred to hereafter as autophagy). One can imagine autophagy, or cellular self-digestion, in its simplest form as single cell's adaptation to starvation: if there is a lack of nutrition in the surroundings, a cell is forced to break down parts of its own reserves to stay alive until the situation improves. During autophagy, portions of the cytosol are engulfed by a membrane sac resulting in a double-membrane vesicle, called autophagosome/autophagic vacuole, which deliver cytoplasmic cargo to lysosomes (Figure 1). After fusion with lysosomes, the

[1]These two authors contributed equally to this work.

The Prion Protein 2010 Jörg Tatzelt (Ed.) Savanna Press, UK
ISBN: 978-0-9543335-2-2. savannapress.com

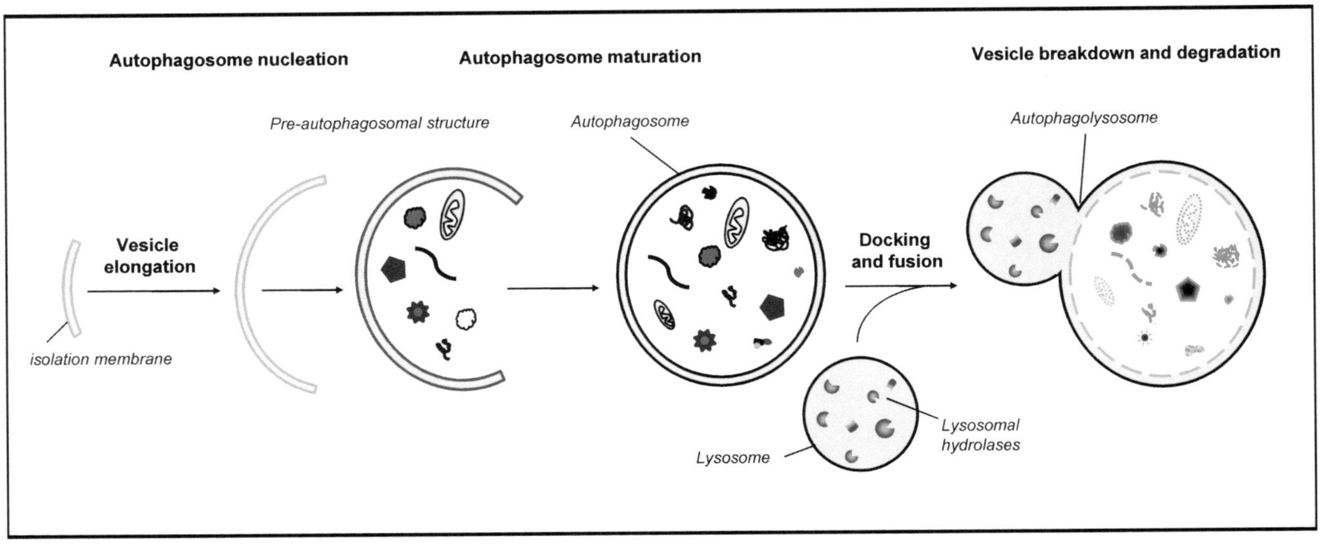

Figure 1. Schematic representation of main cellular steps in macroautophagy.
Autophagosome nucleation starts with an isolation membrane in the cytosol which matures to an autophagosome vesicle. By docking and fusion to lysosomes the content of this autophagolysosome gets access to the lysosomal degradation machinery.

protein and organelle contents of the autophagosome are degraded by acidic lysosomal hydrolases and recycled (Klionsky and Ohsumi, 1999; Yoshimori, 2004) (Figure 1). In single-cell organisms such as yeasts, this starvation response is one of the primary functions of autophagy, but in fact this role extends up to humans. For example, even on a day-to-day basis, autophagy is activated between meals in organs such as the liver to maintain its metabolic functions, supplying amino acids and energy through catabolism (Kuma et al., 2004; Mizushima and Klionsky, 2007).

Regulation and manipulation of autophagy

Autophagy is conserved among eukaryotes and has been characterized from yeast to man (Reggiori and Klionsky, 2002). Basal levels of autophagy are important for maintaining normal cellular homeostasis. As the autophagic process has the capacity for large scale degradation, unregulated degradation of the cytoplasm is likely to be deleterious. Thus, a tight cellular regulation of the autophagic process is important so that it is induced when needed, but otherwise maintained at a basal non-deleterious level.

Several protein kinases regulate autophagy. The best characterized is the mammalian target of rapamycin, mTOR, which is the major inhibitory signal that shuts off autophagy in the presence of growth factors and nutrients (Kamada et al., 2000). Downstream of mTOR, numerous proteins encoded by *Atg* genes are essential for the execution of autophagy (Levine and Klionsky, 2004). Some of the other regulatory molecules that control autophagy include 5′-AMP-activated protein kinase (AMPK), BH3-only proteins, the inositol 1,4,5-trisphosphate receptor (IP_3R), Erk1/2 and calcium (Criollo et al., 2007; Maiuri et al., 2007; Meijer and Codogno, 2006; Rubinsztein et al., 2007). Autophagy can also be pharmacologically induced by inhibiting negative regulators such as mTOR via the compound rapamycin (Rubinsztein et al., 2007) or by mTOR-independent inducers of autophagy such as trehalose and lithium (Sarkar et al., 2007a; Sarkar et al., 2005). Pharmacological inhibitors of autophagy are for instance 3-methyladenine (3-MA), Wortmannin and LY294002 (Blommaart et al., 1997; Seglen and Gordon, 1982).

Although a complete picture of autophagy regulation is not yet available, breakthroughs in yeast genetics and analysis of mammalian homologues of the autophagy-related (Atg) proteins identified in yeast (Harding et al., 1995; Suzuki and Ohsumi, 2007; Tsukada and Ohsumi, 1993) have greatly improved the understanding of autophagy and its regulation. Several aspects of regulation mechanisms have recently been reviewed in great detail (Botti et al., 2006; Gozuacik and Kimchi, 2007; Meijer and Codogno, 2006; Yorimitsu and Klionsky, 2007).

Monitoring autophagy

Basal and induced levels of autophagy are important with regard to its role in human health and disease. With the rapidly advancing research in the autophagy field a range of biochemical and morphological methods has been developed to monitor autophagy [e.g. reviewed in (Klionsky et al., 2007; Mizushima, 2004)]. These methods are useful and reliable to monitor autophagy in yeast, however, there is some confusion regarding some methods to measure autophagy in higher eukaryotes. A key point in monitoring autophagy is that there is a difference between measurements that monitor the numbers of autophagosomes versus those that measure flux through the autophagic pathway. Thus, a block in autophagic flux, for example due to disturbance in lysosomal function, results in autophagosome accumulation which needs to be differentiated from fully functional autophagy that includes delivery to and degradation within lysosomes (in most higher eukaryotes) or the vacuole (in plants and fungi). Recently, a number of studies about the selection and interpretation of the methods that can be used to examine autophagy and related processes were published, recommending the use of multiple assays to verify an autophagic response (Kawai et al., 2006; Klionsky et al., 2008; Mizushima and Yoshimori, 2007).

Autophagy in physiology and patho-physiology

Among recent advances, an exciting finding is the striking pleiotropy of autophagy. Beyond its classical role in nutrient supply under starvation and turnover of organelles and proteins, autophagy contributes to various physiological processes such as intracellular cleansing, differentiation, development, longevity, elimination of invading pathogens and antigen transport to the innate and adaptive immune systems or counteracting endoplasmic reticulum stress and diseases characterized by the accumulation of protein aggregates (Levine and Kroemer, 2008; Levine and Yuan, 2005; Lum et al., 2005; Maiuri et al., 2007; Mizushima et al., 2008; Yorimitsu and Klionsky, 2007). However, in the context of cancer, potentially this pro-survival function seems to be maladaptive (Mathew et al., 2007). This takes us to the other face of autophagy and its connections to patho-physiology and disease. Besides cancer, autophagy plays a role in a number of infectious and inflammatory diseases and in protein 'unfolding and misfolding' diseases that lead to neuronal, muscle and liver degeneration or heart failure (reviewed in Levine and Deretic, 2007; Levine and Kroemer, 2008; Mizushima et al., 2008).

With respect to the importance of tight regulation of autophagy, perhaps the most fundamental point is that either too little or too much autophagy can be deleterious, a complex balance resulting in its dual role in survival and adaptation or cell death. However, in response to most forms of cellular stress, autophagy plays a cytoprotective role, because *Atg* gene knockdown/knockout accelerates rather than delays cell death (Levine and Yuan, 2005; Maiuri et al., 2007). Within the cell death research field, autophagy has long been defined as a form of non-apoptotic, or type II, programmed cell death (Clarke, 1990; Kovacs et al., 1986). However, due to the recent findings a consensus is emerging that autophagy might be a cell death impostor which, in reality, functions primarily to promote cellular and organismal health (Kroemer and Levine, 2008).

Autophagy and neurodegenerative diseases

As mentioned above, autophagy occurs at basal, constitutive levels in the cell. Recent studies have highlighted the importance of basal autophagy in intracellular quality control. The demand for basal autophagy is tissue-specific, though. In liver and other tissues where cells, such as neurons and myocytes, do not divide after differentiation basal autophagy is of great relevance (Hara et al., 2006; Komatsu et al., 2006; Komatsu et al., 2005; Komatsu et al., 2007b; Nakai et al., 2007). Several studies suggest a crucial role of autophagy in neurodegenerative diseases, including Alzheimer's disease, Parkinson's disease, tauopathies and polyglutamine expansion diseases like Huntingon's disease (Berger et al., 2006; Iwata et al., 2005; Mizushima and Hara, 2006; Nixon et al., 2005; Qin et al., 2003; Ravikumar et al., 2002; Rubinsztein, 2006; Rubinsztein et al., 2005; Ventruti and Cuervo, 2007; Webb et al., 2003). A number of *in vivo* studies during the last years showed that conventional autophagy knockout mice die during embryogenesis or the neonatal period (Fimia et al., 2007; Komatsu et al., 2005; Kuma et al., 2004; Qu et al., 2003; Takahashi et al., 2007; Yue et al., 2003). Mice with neural-tissue-specific knockouts of these genes survive the postnatal starvation period. However, these mice develop progressive motor deficits and display abnormal reflexes, and ubiquitin-positive inclusion bodies accumulate in their neurons (Hara et al., 2006; Komatsu et al., 2006). Studies showed that the CNS, in contrast to other organ systems, displays only low levels of autophagosomes under normal conditions and even after starvation, but it was also demonstrated that constitutive turnover of cytosolic contents by autophagy is indispensable, even in the absence of expression of any disease-associated mutant proteins (Mizushima et al., 2004; Nixon et al., 2005).

Despite the important function of basal autophagy in healthy individuals, the requirement for autophagy is even more evident under disease conditions and levels of autophagosomes can be dramatically increased in injured or degenerating neurons (Petersen et al., 2001). Available data state, beyond any doubt, that autophagy has a beneficial effect of protecting against neurodegeneration. There are several hypotheses about how autophagy can prevent neurodegeneration. However, this is not yet fully understood. One idea is that autophagy eliminates aggregated and aggregate-prone proteins (Bjorkoy et al., 2005; Iwata et al., 2005; Komatsu et al., 2007a; Pankiv et al., 2007; Ravikumar et al., 2002). Concerning Alzheimer's disease and the involvement of autophagy, another hypothesis is that impaired autophagic flux provides a novel site for Aβ peptide production (Yu et al., 2005). Thus, it is reasonable to assume that autophagy could be a therapeutic target for treatment of these neurodegenerative diseases because of its protective role (Rubinsztein et al., 2007).

Recent detailed studies underlined that degradation of disease-related mutant proteins is highly dependent on autophagy, in addition to the ubiquitin–proteasome system. Examples include studies performed with extended polyglutamine-containing proteins that cause various neurodegenerative diseases (Martinez-Vicente and Cuervo, 2007; Rubinsztein, 2006; Rubinsztein et al., 2005). It was shown that the clearance of aggregate-prone proteins, such as mutant huntingtin fragments or mutant forms of α-synuclein causing Huntington´s and Parkinson´s disease, respectively, can be mediated by autophagy (Ravikumar et al., 2002; Webb et al., 2003). Animal models of Huntington´s disease and of other proteinopathies revealed that treatment with rapamycin, a known inducer of autophagy, accelerates the clearance of toxic proteins (Berger et al., 2006; Iwata et al., 2005; Qin et al., 2003; Ravikumar et al., 2004; Shibata et al., 2006). Induction of autophagy, mediated by lithium and trehalose, has been seen to accelerate the clearance of mutant huntingtin and α-synucleins (Sarkar et al., 2007a; Sarkar et al., 2005). The beneficial effect of up-regulated autophagy has also been described for other diseases associated with aggregate-prone proteins, such as Alzheimer´s disease, forms of motor neuron disease caused by mutations in superoxide dismutase 1 (SOD1), and forms of peripheral neuropathy caused by mutations in peripheral myelin protein 22 (PMP22) (Berger et al., 2006; Fortun et al., 2003; Kabuta et al., 2006).

Moreover, up-regulation of autophagy by regulatory protein kinase complex Target of Rapamycin (TOR) inhibitors such as rapamycin and its analogue CCI-779 protects against neurodegeneration seen in polyglutamine disease models in Drosophila and mice (Ravikumar et al., 2004). Recently, small-molecule enhancers of rapamycin were identified (Sarkar et al., 2007b), which improve the clearance of mutant huntingtin and α-synuclein and protect against neurodegeneration in a Drosophila Huntington's disease model. Of note, the effects of small molecule enhancers of rapamycin are independent of TOR, which adds to the possibility of using them in combination with rapamycin for therapeutic purposes. However, it is important to put into consideration that the potential lying in manipulating autophagy as a therapeutic approach for neurodegenerative diseases strongly depends on complex disease-specific factors. On the cellular level, a profound understanding of how the autophagic machinery is involved during the pathogenic course of the disease will be relevant. On the whole organism level, the impact of autophagy manipulation will depend on the dynamics of the disease in terms of dissemination and the changes that occur in the organism.

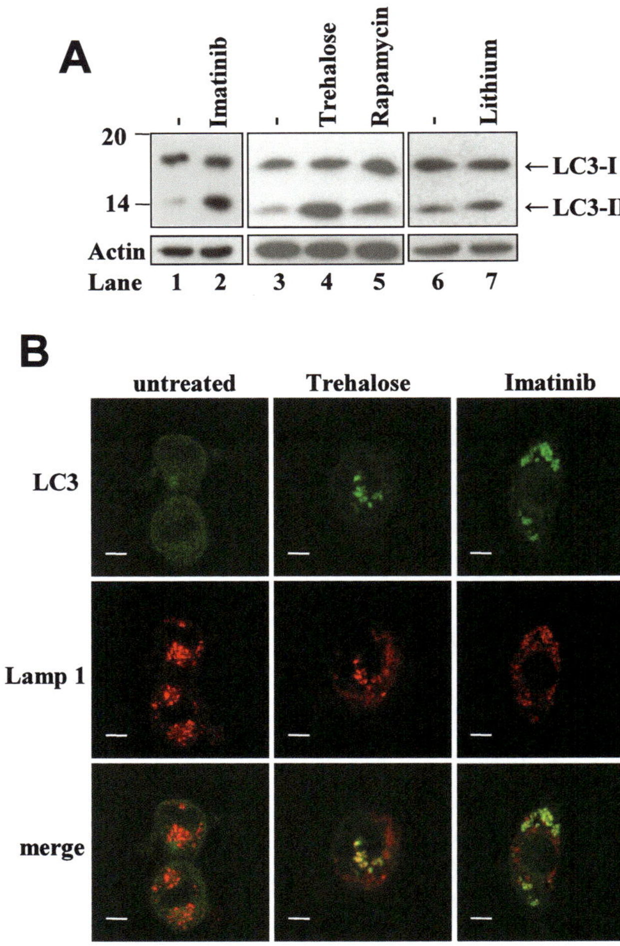

Figure 2. Monitoring autophagosome formation.
(A) ScN2a cells were treated with 10 µM imatinib, 100 mM trehalose, or 200 nM rapamycin for 48 h, or 10 mM lithium for 24 h. Mock-treated cells served as controls. Subsequently, cell lysates were probed in SDS-PAGE using anti-LC3 antibody. Higher amounts of LC3-II are detected in compound-treated cells compared to mock-treated controls, indicating induction of autophagy and autophagosome formation.
(B) ScN2a cells were transfected with GFP-LC3 and either left untreated, treated with 100 mM trehalose or 10µM imatinib for 48 h. Cells were stained with anti-lamp 1 and cy2-conjugated secondary antibody and subsequently analyzed by confocal microscopy. Both autophagosome formation (upper panels) and co-localization of autophagosomes and lysosomes (lower panels) is observed in compound-treated cells.

Autophagy and prion infection

Autophagic vacuoles were described in neurons in experimental models of prion disease in mice and hamsters (Boellaard et al., 1991; Boellaard et al., 1989). In addition, the appearance of multi-vesicular bodies and autophagic vacuoles was observed in prion-infected cultured neuronal cells (Schatzl et al., 1997). More recently, it was found that autophagic vacuoles are formed in neuronal perikarya, neurites and synapses in experimentally induced scrapie, Creutzfeldt-Jakob disease (CJD) and Gerstmann-Sträussler-Scheinker (GSS) syndrome (Liberski et al., 2004) and autophagic vacuoles were identified in synapses in various forms of human prion disease (Sikorska et al., 2004).

Another interesting correlation between prion diseases and autophagy was observed in studies on *scrg1* (scrapie responsive gene 1). This gene was up-regulated in brains of scrapie prion- and BSE-infected mice and in brains of patients with sporadic CJD (Dandoy-Dron et al., 2000; Dandoy-Dron et al., 1998; Dron et al., 1998). In the CNS of prion-infected mice up-regulated Scrg1 was associated with autophagic vacuoles which were observed at the terminal stage of disease (Dron et al., 2005). Consequently, it was suggested that Scrg1 might be useful as a marker for neuronal autophagy in prion diseases (Dron et al., 2006).

With regard to the PrP homologue doppel (Dpl), it was shown that ectopic expression of Dpl in CNS neurons of prion protein knockout-mice (*Ngsk*; $NP^{0/0}$) results in late-onset ataxia due to extensive Purkinje cell (PC) death (Moore et al., 1999; Rossi et al., 2001; Sakaguchi et al., 1996). In line with this, it was demonstrated that preceding and during such PC loss the protein levels of both Scrg1 and the well established autophagic markers LC3-II and p62 were increased, whereas mRNA expression levels were stable (Heitz et al., 2008). It was suggested that CNS expression of Dpl might trigger autophagy and that the apoptotic cascade might be triggered by a progressive dysfunction of autophagy.

Besides such descriptions of autophagy in prion disease models, the putative involvement of PrP^c in autophagic pathways was recently described. An increased expression of LC3-II was observed in hippocampal neurons of Zürich I *Prnp*$^{-/-}$ mice as compared to wild-type control neurons under serum deprivation and this up-regulation was counter-acted by reintroduction of PrP^c into *Prnp*$^{-/-}$ cells (Oh et al., 2008). As such counter-regulation was not detectable for PrP^c lacking the octapeptide region, it was suggested that the octapeptide region of PrP^c may play a crucial role in control of autophagy in neuronal cells as mediated by PrP^c.

Concerning again the role of autophagy in prion disease, it was proposed that autophagy may contribute to formation of spongiform changes, a pathological hallmark in prion-affected brains, and may be activated by apoptosis (Liberski et al., 2008; Liberski et al., 2002; Liberski et al., 2004). In contrast to this assumption that autophagy plays a disease-promoting role, it is also quite conceivable that the observed increase in autophagic vacuoles in prion disease models is due to activation of the autophagic machinery as a defense mechanism, leading even to degradation of prions. Support for such a protective role of autophagy in prion disease was described in studies addressing a member of the galactin family of proteins, namely galactin-3. Reduced levels of the lysosomal activation marker LAMP-2 were observed in prion-infected galactin-$3^{-/-}$-mice and, interestingly, in brain tissue of prion-infected wild-type and galactin-$3^{-/-}$-mice, lower mRNA levels of autophagy markers Beclin-1 and Atg5 were detected as compared to mock-infected control brains (Mok et al., 2007). Therefore, the authors suggested that endosomal/lysosomal dysfunction in combination with reduced autophagy may contribute to development of prion disease.

Previously, our group showed that imatinib, a drug used to treat chronic myelogenous leukemia, is activating lysosomal degradation of PrP^{Sc} (Ertmer et al., 2004) and is at the same time a potent inducer of autophagy and/or autophagosome formation (Ertmer et al., 2007). In prion-infected mice, imatinib treatment at an early phase of peripheral infection delayed both the neuroinvasion of PrP^{Sc} and the onset of clinical disease (Yun et al., 2007). Unfortunately, drug application at time points when neuroinvasion was already accomplished provoked no clear PrP^{Sc} clearance effects in CNS, probably due to ineffective blood-brain barrier crossing of the drug. The beneficial effect of up-regulated autophagy was shown for several neurodegenerative diseases associated with aggregate-prone proteins (Berger et al., 2006; Fortun et al., 2003; Kabuta et al., 2006; Ravikumar et al., 2002; Webb et al., 2003). In line with studies on mutant forms of huntingtin or α-synuclein (Sarkar et al., 2007a; Sarkar et al., 2005) we could recently show that both lithium and trehalose enhance the clearance of PrP^{Sc} in prion-infected cells by induction of autophagy (Aguib et al., 2009; Heiseke et al., 2009). As mentioned, a common method to monitor autophagosome formation is to analyse the level of LC3-II which is associated with autophagosome membranes (Kabeya et al., 2000). Chemical compounds used in our studies all increased the amount of LC3-II in prion-infected cells in immunoblot analysis (Figure 2A). LC3 fused to green fluorescent protein (GFP) is another tool to measure induction of autophagosome formation. Prion-infected neuroblastoma (ScN2a) cells treated with autophagy inducing drugs exhibited punctuate GFP staining (exemplarily shown in Figure 2B for trehalose and imatinib), indicating the association of GFP-LC3 with autophagosomal membranes as a result of induction of autophagy.

In our studies we provided the first direct evidence that induction of autophagy results in degradation of cellular PrP^{Sc}. Inhibition of autophagy by pharmacological interference and siRNA gene-silencing of essential members of the autophagy machinery impaired the capacity of compound-induced autophagy in reducing cellular levels of PrP^{Sc}. Of note, as induced autophagy was able to degrade aggregate-prone proteins accumulating within endosomal/lysosomal vesicles, as is the case for PrP^{Sc}, we show that autophagy is not only an important clearance route for cytosolic aggregate-prone proteins. Besides compounds inducing autophagy in an mTOR-independent manner (e.g. lithium, trehalose), we studied rapamycin, a drug widely used to activate autophagy by inhibiting mTOR. Rapamycin also reduced the level of cellular PrP^{Sc}, showing that both autophagy inducing pathways, mTOR-dependent and -independent, may be involved in the degradation of PrP^{Sc}. Reduction of PrP^{Sc} in prion-infected ScN2a cells upon treatment with several autophagy inducing drugs is shown in Figure 3.

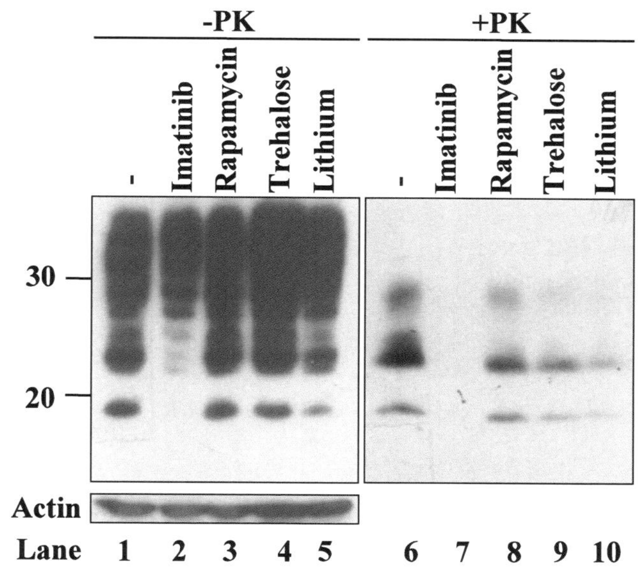

Figure 3. Reduction of PrP^{Sc} upon treatment of cells with different autophagy inducing compounds.
ScN2a cells were either left untreated or treated with 10 μM imatinib, 200 nM rapamycin, 100 mM trehalose, or 10 mM lithium for 48 h and cell lysates subsequently analyzed by SDS-PAGE using anti-PrP monoclonal antibody 4H11. Upon PK digestion (lanes 6-10) less PrP^{Sc} is observed in compound-treated cells compared to untreated control cells.

PrP^c localizes via a glycosylphosphatidylinositol (GPI)-anchor at the outer leaflet of the plasma membrane in cholesterol- and sphingolipid-rich microdomains (Taraboulos et al., 1995) and can move laterally to detergent-soluble domains within the plasma-membrane for subsequent internalization (Sunyach et al., 2003). Therefore, it is possible that upon internalization PrP^c is in reach of the autophagic degradation machinery. We showed that lithium is not only reducing PrP^{Sc} but also levels of PrP^c in an autophagy-dependent manner. We observed no reduction of PrP^c in lithium-treated, autophagy-deficient fibroblasts whereas the level of PrP^c was slightly but significantly reduced in wild-type fibroblasts upon lithium treatment. In the recent past it has been shown that reduction of PrP^c by shedding of the protein from the membrane or by down-regulation of PrP^c reduces conversion of PrP^c into its pathogenic isoform PrP^{Sc} by limiting the amount of PrP^c substrate available for conversion (Aguib et al., 2008; Heiseke et al., 2008; Marella et al., 2002; Parkin et al., 2004). Therefore, reduction of PrP^c by lithium-induced autophagy may indirectly contribute to reduction of PrP^{Sc} by autophagy. As we did not observe reduced levels of PrP^c upon treatment of cells with other autophagy inducing compounds this phenomenon seems to be compound-specific and the exact molecular mechanism remains to be deciphered.

To test whether autophagy-inducing compounds are candidates for therapeutic approaches against prion infection we treated intraperitoneally prion-infected mice

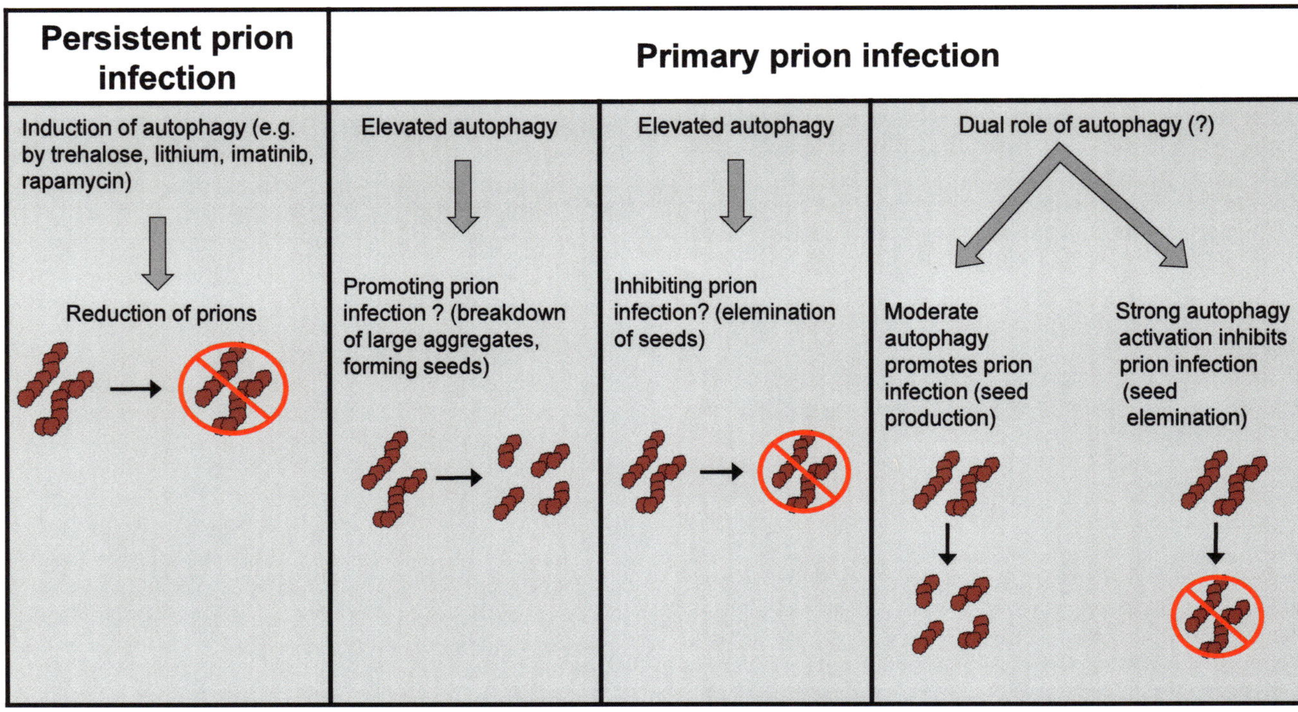

Figure 4. Model of established and putative actions of autophagy in prion infection scenarios.

(Aguib et al., 2009; Heiseke et al., 2009). Rapamycin treatment of prion-infected mice initiated in the last third of incubation time (i.e. day 100 p.i.), mimicking a pre-clinical therapeutic situation, showed a significant prolongation of prion incubation times as compared to mock-treated control mice. Similar findings were obtained with lithium, although less pronounced (Heiseke et al., 2009). Trehalose treatment did not prolong incubation times, but clearly showed effects on the appearance of PrP^{Sc} in spleens (Aguib et al., 2009). Depending on when trehalose treatment was started, peripheral accumulation of PrP^{Sc} was delayed. As was the case with imatinib treatment, this probably also reflects that the process of neuroinvasion was decelerated. Overall, although still preliminary, these *in vivo* studies strongly indicate that autophagy inducing compounds are beneficial in prion disease scenarios and ask for further and more complex studies, including also combination of drugs.

More than one role of autophagy in prion infection?

It is conceivable that the basic process of autophagy has a physiological role in prion infection and might be used by cells for controlling or counter-acting cellular prion infection. In line with this assumption, we observed an increased amount of PrP^{Sc} in cells in which autophagy was reduced by treatment with autophagy inhibitors (Aguib et al., 2009). The increased PrP^{Sc} level may result in an increased conversion of PrP^{c} into PrP^{Sc}, subsequently leading to more cellular PrP^{Sc}. In addition, when we assayed for altered regulation of autophagy in primary infection models, a correlation of prion infection and increase in autophagy was obtained (Heiseke, Aguib and Schatzl, personal communication). Manipulation of autophagy in primary prion infection, either increasing or decreasing autophagy, also had reproducible effects on primary prion infection.

Besides more detailed *in vitro* studies appropriate *in vivo* experiments are required to elucidate the impact of autophagy on prion infection and to validate whether autophagy plays a general role in prion disease scenarios. One possibility is using mice expressing GFP-LC3, allowing direct correlation analysis of hallmarks of prion infection and autophagy *in vivo* (Baier, Aguib, Heiseke and Schatzl, personal communication). As presently available mice with a neuron-specific conditional knockout of *Atg5* or *Atg7* die soon after birth at time points which make them not accessible for classical prion infection studies, alternative Cre deleter mice have to be crossed in, in order to generate a neuron-specific and postnatal knockout of these genes. This would enable to directly assess the impact of autophagy in prion incubation time and prion disease in *in vivo* models, due to increased life-span of neuron-specific *Atg* knockout mice.

Although there is now good evidence from our work that induction of autophagy is beneficial in prion-infected cells and animals, it is not clear whether increased or decreased autophagy also can have deleterious effects. One scenario would be that cells impaired in autophagy might be more susceptible to prion infection as they are lacking a putative defence mechanism. In primary infection models we started to test whether increase or decrease of basal autophagy is a modifier of prion infection and susceptibility to prion infection. Another possibility worth to envision is that autophagy also might be a positive factor for prion propagation. A moderate basal level of cellular autophagy, that is likely to be present even in treatment situations with autophagy inhibitors, might be beneficial for generating smaller PrP^{Sc} seeds, which are known to be more efficient templates for conversion of PrP^{c} into PrP^{Sc} than are larger aggregates (Silveira et al., 2005). Hence, although induced autophagy can reduce or clear prions and PrP^{Sc} seeds, a moderate level of autophagy might support PrP^{Sc} seed production at certain stages of prion infection, thereby promoting prion disease. Experimental evidence for this was obtained when we tried to infect fibroblasts from $Atg5^{-/-}$ mice with murine prions (Heiseke, Aguib and Schatzl, personal communication). Whereas wild-type fibroblasts could readily be infected with prions, knockout fibroblasts were significantly less infectable. Re-introduction of *Atg5* via lentiviral transduction clearly improved this deficiency in primary prion infection (Heiseke, Aguib and Schatzl, personal communication). This provides reliable evidence that a certain level of autophagy can be both a positive modifier of prion infection and a supporting factor, which in certain scenarios is even essential for prion propagation. As outlined in Figure 4, a combination of anti-prion and prion-promoting effects of autophagy cannot be excluded as well. This remains us with a scenario reminding of the seeming paradox of the role of autophagy in cell death and cancer biology (Mizushima et al., 2008).

Conclusions and future directions

The systematic analysis of autophagy in prion infection scenarios is still rather incomplete. From what is found so far there is good experimental evidence from *in vitro* and partly also *in vivo* studies that induction of autophagy, e.g. by chemical compounds, can clearly have beneficial effects on prion infection. The cellular load of PrP^{Sc} and prion infectivity is reduced, most probably by an increase in lysosomal degradation, shifting the equilibrium between prion propagation and clearance towards the latter. The exact molecular mechanisms are still incompletely understood, in particular as the very vast majority of PrP^{Sc}/prions reside within endosomal and lysosomal vesicles. On the other hand this clearly shows that not only cytosolic materials are prone to autophagic degradation. In future work it has to be studied how prion propagation, prion trafficking and re-cycling, and finally prion clearance are interconnected with the autophagic pathway. In addition, more work is needed to elucidate whether induction of autophagy in this context has to be mTOR-dependent, -independent, or both. Finally, bringing this knowledge to translational research and bridging therapeutic anti-prion concepts will be extremely difficult, as it is with basically all other anti-prion strategies so far. Besides suboptimal pharmacokinetics and possible side effects the ultralarge obstacle rests with ineffective passing of the blood-brain-barrier. This does not exclude effects in post-exposure situations in peripheral prion infection scenarios and a sophisticated combination of compounds which target divers 'anti-prion' pathways still is a reasonable goal.

The biological function of autophagy *per se* in prion infection and disease is still searching answers. Preliminary data indicate that the cellular level of autophagy can be a modifier of susceptibility to prion infection, although at present it is still difficult to dissect whether changes in autophagy are a pre-requisite or consequence of cellular prion infection. Questions like this are readily accessible in *in vitro* studies. As mentioned, autophagy might also have a dual function and there might be situations in which autophagy is needed for prion propagation. Work from yeast prions and nowadays from mammalian cell culture systems for studying prion-like properties (Krammer et al., 2009) indicates that the kinetics of aggregate formation needs breaking-up of aggregates and fibrils, probably involving disaggregase activities. It will be interesting to study whether autophagy is a cellular mechanism involved in this scenario.

Another challenge will be to establish reliable *in vivo* models for studying prion infection and autophagy side by side. Faced with early lethality in *Atg* knockout mice, one way to go might be crossing the available conditional knockout mice, which are neuron-specifically floxed, with alternative Cre deleter mice to gain both postnatal knockout and prolonged life time which then allows performing standard prion incubation time assays.

In summary, there seems to be a fascinating interplay between prion infections and autophagy. Although extensive future studies will be necessary, there is a high probability that it is both worth and feasible to decipher their mutual interaction at a molecular level.

Acknowledgements

This work was supported by SFB-596 (project A8), APRI Canada, and was performed within the framework of European Union FP6 Network of Excellence 'Neuroprion'. Y.A. was supported by grants from Technische Universität München (TUM) and the German Academic Exchange Service (DAAD).

References

Aguib, Y., Gilch, S., Krammer, C., Ertmer, A., Groschup, M.H., and Schatzl, H.M. (2008). Neuroendocrine cultured cells counteract persistent prion infection by down-regulation of PrPc. Molecular and cellular neurosciences *38*, 98-109.

Aguib, Y., Heiseke, A., Gilch, S., Riemer, C., Baier, M., Schatzl, H.M., and Ertmer, A. (2009). Autophagy induction by trehalose counteracts cellular prion infection. Autophagy *5*, 361-369.

Aguzzi, A., and Haass, C. (2003). Games played by rogue proteins in prion disorders and Alzheimer's disease. Science (New York, NY *302*, 814-818.

Aguzzi, A., and Polymenidou, M. (2004). Mammalian prion biology: one century of evolving concepts. Cell *116*, 313-327.

Berger, Z., Ravikumar, B., Menzies, F.M., Oroz, L.G., Underwood, B.R., Pangalos, M.N., Schmitt, I., Wullner, U., Evert, B.O., O'Kane, C.J., *et al.* (2006). Rapamycin alleviates toxicity of different aggregate-prone proteins. Human molecular genetics *15*, 433-442.

Bjorkoy, G., Lamark, T., Brech, A., Outzen, H., Perander, M., Overvatn, A., Stenmark, H., and Johansen, T. (2005). p62/SQSTM1 forms protein aggregates degraded by autophagy and has a protective effect on huntingtin-induced cell death. The Journal of cell biology *171*, 603-614.

Blommaart, E.F., Krause, U., Schellens, J.P., Vreeling-Sindelarova, H., and Meijer, A.J. (1997). The phosphatidylinositol 3-kinase inhibitors wortmannin and LY294002 inhibit autophagy in isolated rat hepatocytes. European journal of biochemistry / FEBS *243*, 240-246.

Boellaard, J.W., Kao, M., Schlote, W., and Diringer, H. (1991). Neuronal autophagy in experimental scrapie. Acta neuropathologica *82*, 225-228.

Boellaard, J.W., Schlote, W., and Tateishi, J. (1989). Neuronal autophagy in experimental Creutzfeldt-Jakob's disease. Acta neuropathologica *78*, 410-418.

Botti, J., Djavaheri-Mergny, M., Pilatte, Y., and Codogno, P. (2006). Autophagy signaling and the cogwheels of cancer. Autophagy *2*, 67-73.

Bursch, W., and Ellinger, A. (2005). Autophagy--a basic mechanism and a potential role for neurodegeneration. Folia neuropathologica / Association of Polish Neuropathologists and Medical Research Centre, Polish Academy of Sciences *43*, 297-310.

Clarke, P.G. (1990). Developmental cell death: morphological diversity and multiple mechanisms. Anatomy and embryology *181*, 195-213.

Cohen, F.E., Pan, K.M., Huang, Z., Baldwin, M., Fletterick, R.J., and Prusiner, S.B. (1994). Structural clues to prion replication. Science (New York, NY *264*, 530-531.

Criollo, A., Maiuri, M.C., Tasdemir, E., Vitale, I., Fiebig, A.A., Andrews, D., Molgo, J., Diaz, J., Lavandero, S., Harper, F., *et al.* (2007). Regulation of autophagy by the inositol trisphosphate receptor. Cell death and differentiation *14*, 1029-1039.

Dandoy-Dron, F., Benboudjema, L., Guillo, F., Jaegly, A., Jasmin, C., Dormont, D., Tovey, M.G., and Dron, M. (2000). Enhanced levels of scrapie responsive gene mRNA in BSE-infected mouse brain. Brain research *76*, 173-179.

Dandoy-Dron, F., Guillo, F., Benboudjema, L., Deslys, J.P., Lasmezas, C., Dormont, D., Tovey, M.G., and Dron, M. (1998). Gene expression in scrapie. Cloning of a new scrapie-responsive gene and the identification of increased levels of seven other mRNA transcripts. The Journal of biological chemistry *273*, 7691-7697.

Dron, M., Bailly, Y., Beringue, V., Haeberle, A.M., Griffond, B., Risold, P.Y., Tovey, M.G., Laude, H., and Dandoy-Dron, F. (2005). Scrg1 is induced in TSE and brain injuries, and associated with autophagy. The European journal of neuroscience *22*, 133-146.

Dron, M., Bailly, Y., Beringue, V., Haeberle, A.M., Griffond, B., Risold, P.Y., Tovey, M.G., Laude, H., and Dandoy-Dron, F. (2006). SCRG1, a potential marker of autophagy in transmissible spongiform encephalopathies. Autophagy *2*, 58-60.

Dron, M., Dandoy-Dron, F., Guillo, F., Benboudjema, L., Hauw, J.J., Lebon, P., Dormont, D., and Tovey, M.G. (1998). Characterization of the human analogue of a Scrapie-responsive gene. The Journal of biological chemistry *273*, 18015-18018.

Ertmer, A., Gilch, S., Yun, S.W., Flechsig, E., Klebl, B., Stein-Gerlach, M., Klein, M.A., and Schatzl, H.M. (2004). The tyrosine kinase inhibitor STI571 induces cellular clearance of PrPSc in prion-infected cells. The Journal of biological chemistry *279*, 41918-41927.

Ertmer, A., Huber, V., Gilch, S., Yoshimori, T., Erfle, V., Duyster, J., Elsasser, H.P., and Schatzl, H.M. (2007). The anticancer drug imatinib induces cellular autophagy. Leukemia *21*, 936-942.

Fimia, G.M., Stoykova, A., Romagnoli, A., Giunta, L., Di Bartolomeo, S., Nardacci, R., Corazzari, M., Fuoco, C., Ucar, A., Schwartz, P., *et al.* (2007). Ambra1 regulates autophagy and development of the nervous system. Nature *447*, 1121-1125.

Fortun, J., Dunn, W.A., Jr., Joy, S., Li, J., and Notterpek, L. (2003). Emerging role for autophagy in the removal of aggresomes in Schwann cells. J Neurosci *23*, 10672-10680.

Gozuacik, D., and Kimchi, A. (2007). Autophagy and cell death. Current topics in developmental biology *78*, 217-245.

Hara, T., Nakamura, K., Matsui, M., Yamamoto, A., Nakahara, Y., Suzuki-Migishima, R., Yokoyama, M., Mishima, K., Saito, I., Okano, H., *et al.* (2006). Suppression of basal autophagy in neural cells causes neurodegenerative disease in mice. Nature *441*, 885-889.

Harding, T.M., Morano, K.A., Scott, S.V., and Klionsky, D.J. (1995). Isolation and characterization of yeast mutants in the cytoplasm to vacuole protein targeting pathway. The Journal of cell biology *131*, 591-602.

Heiseke, A., Aguib, Y., Riemer, C., Baier, M., and Schatzl, H.M. (2009). Lithium induces clearance of protease resistant prion protein in prion-infected cells by induction of autophagy. Journal of neurochemistry *109*, 25-34.

Heiseke, A., Schobel, S., Lichtenthaler, S.F., Vorberg, I., Groschup, M.H., Kretzschmar, H., Schatzl, H.M., and Nunziante, M. (2008). The novel sorting nexin SNX33 interferes with cellular PrP formation by modulation of PrP shedding. Traffic *9*, 1116-1129.

Heitz, S., Grant, N.J., Leschiera, R., Haeberle, A.M., Demais, V., Bombarde, G., and Bailly, Y. (2008). Autophagy and Cell Death of Purkinje Cells Overexpressing Doppel in Ngsk Prnp-deficient Mice. Brain pathology (Zurich, Switzerland).

Iwata, A., Riley, B.E., Johnston, J.A., and Kopito, R.R. (2005). HDAC6 and microtubules are required for autophagic degradation of aggregated huntingtin. The Journal of biological chemistry *280*, 40282-40292.

Kabeya, Y., Mizushima, N., Ueno, T., Yamamoto, A., Kirisako, T., Noda, T., Kominami, E., Ohsumi, Y., and Yoshimori, T. (2000). LC3, a mammalian homologue of yeast Apg8p, is localized in autophagosome membranes after processing. The EMBO journal *19*, 5720-5728.

Kabuta, T., Suzuki, Y., and Wada, K. (2006). Degradation of amyotrophic lateral sclerosis-linked mutant Cu,Zn-superoxide dismutase proteins by macroautophagy and the proteasome. The Journal of biological chemistry *281*, 30524-30533.

Kamada, Y., Funakoshi, T., Shintani, T., Nagano, K., Ohsumi, M., and Ohsumi, Y. (2000). Tor-mediated induction of autophagy via an Apg1 protein kinase complex. The Journal of cell biology *150*, 1507-1513.

Kawai, A., Takano, S., Nakamura, N., and Ohkuma, S. (2006). Quantitative monitoring of autophagic degradation. Biochemical and biophysical research communications *351*, 71-77.

Klionsky, D.J., Abeliovich, H., Agostinis, P., Agrawal, D.K., Aliev, G., Askew, D.S., Baba, M., Baehrecke, E.H., Bahr, B.A., Ballabio, A., *et al.* (2008). Guidelines for the use and interpretation of assays for monitoring autophagy in higher eukaryotes. Autophagy *4*, 151-175.

Klionsky, D.J., Cuervo, A.M., and Seglen, P.O. (2007). Methods for monitoring autophagy from yeast to human. Autophagy *3*, 181-206.

Klionsky, D.J., and Ohsumi, Y. (1999). Vacuolar import of proteins and organelles from the cytoplasm. Annual review of cell and developmental biology *15*, 1-32.

Komatsu, M., Ueno, T., Waguri, S., Uchiyama, Y., Kominami, E., and Tanaka, K. (2007a). Constitutive autophagy: vital role in clearance of unfavorable proteins in neurons. Cell death and differentiation *14*, 887-894.

Komatsu, M., Waguri, S., Chiba, T., Murata, S., Iwata, J., Tanida, I., Ueno, T., Koike, M., Uchiyama, Y., Kominami, E., *et al.* (2006). Loss of autophagy in the central nervous system causes neurodegeneration in mice. Nature *441*, 880-884.

Komatsu, M., Waguri, S., Ueno, T., Iwata, J., Murata, S., Tanida, I., Ezaki, J., Mizushima, N., Ohsumi, Y., Uchiyama, Y., *et al.* (2005). Impairment of starvation-induced and constitutive autophagy in Atg7-deficient mice. The Journal of cell biology *169*, 425-434.

Komatsu, M., Wang, Q.J., Holstein, G.R., Friedrich, V.L., Jr., Iwata, J., Kominami, E., Chait, B.T., Tanaka, K., and Yue, Z. (2007b). Essential role for autophagy protein Atg7 in the maintenance of axonal homeostasis and the prevention of axonal degeneration. Proceedings of the National Academy of Sciences of the United States of America *104*, 14489-14494.

Kovacs, J., Fellinger, E., Karpati, P.A., Kovacs, A.L., and Laszlo, L. (1986). The turnover of autophagic vacuoles: evaluation by quantitative electron microscopy. Biomedica biochimica acta *45*, 1543-1547.

Krammer, C., Kryndushkin, D., Suhre, M.H., Kremmer, E., Hofmann, A., Pfeifer, A., Scheibel, T., Wickner, R.B., Schatzl, H.M., and Vorberg, I. (2009). The yeast Sup35NM domain propagates as a prion in mammalian cells. Proceedings of the National Academy of Sciences of the United States of America *106*, 462-467.

Kroemer, G., and Levine, B. (2008). Autophagic cell death: the story of a misnomer. Nat Rev Mol Cell Biol *9*, 1004-1010.

Kuma, A., Hatano, M., Matsui, M., Yamamoto, A., Nakaya, H., Yoshimori, T., Ohsumi, Y., Tokuhisa, T., and Mizushima, N. (2004). The role of autophagy during the early neonatal starvation period. Nature *432*, 1032-1036.

Levine, B., and Deretic, V. (2007). Unveiling the roles of autophagy in innate and adaptive immunity. Nat Rev Immunol *7*, 767-777.

Levine, B., and Klionsky, D.J. (2004). Development by self-digestion: molecular mechanisms and biological functions of autophagy. Developmental cell *6*, 463-477.

Levine, B., and Kroemer, G. (2008). Autophagy in the pathogenesis of disease. Cell *132*, 27-42.

Levine, B., and Yuan, J. (2005). Autophagy in cell death: an innocent convict? The Journal of clinical investigation *115*, 2679-2688.

Liberski, P.P., Brown, D.R., Sikorska, B., Caughey, B., and Brown, P. (2008). Cell death and autophagy in prion diseases (transmissible spongiform encephalopathies). Folia neuropathologica / Association of Polish Neuropathologists and Medical Research Centre, Polish Academy of Sciences *46*, 1-25.

Liberski, P.P., Gajdusek, D.C., and Brown, P. (2002). How do neurons degenerate in prion diseases or transmissible spongiform encephalopathies (TSEs): neuronal autophagy revisited. Acta neurobiologiae experimentalis *62*, 141-147.

Liberski, P.P., Sikorska, B., Bratosiewicz-Wasik, J., Gajdusek, D.C., and Brown, P. (2004). Neuronal cell death in transmissible spongiform encephalopathies (prion diseases) revisited: from apoptosis to autophagy. The international journal of biochemistry & cell biology *36*, 2473-2490.

Lum, J.J., DeBerardinis, R.J., and Thompson, C.B. (2005). Autophagy in metazoans: cell survival in the land of plenty. Nat Rev Mol Cell Biol *6*, 439-448.

Maiuri, M.C., Zalckvar, E., Kimchi, A., and Kroemer, G. (2007). Self-eating and self-killing: crosstalk between autophagy and apoptosis. Nat Rev Mol Cell Biol *8*, 741-752.

Marella, M., Lehmann, S., Grassi, J., and Chabry, J. (2002). Filipin prevents pathological prion protein accumulation by reducing endocytosis and inducing cellular PrP release. The Journal of biological chemistry *277*, 25457-25464.

Martinez-Vicente, M., and Cuervo, A.M. (2007). Autophagy and neurodegeneration: when the cleaning crew goes on strike. Lancet neurology *6*, 352-361.

Mathew, R., Karantza-Wadsworth, V., and White, E. (2007). Role of autophagy in cancer. Nat Rev Cancer *7*, 961-967.

Meijer, A.J., and Codogno, P. (2006). Signalling and autophagy regulation in health, aging and disease. Molecular aspects of medicine *27*, 411-425.

Mizushima, N. (2004). Methods for monitoring autophagy. The international journal of biochemistry & cell biology *36*, 2491-2502.

Mizushima, N., and Hara, T. (2006). Intracellular quality control by autophagy: how does autophagy prevent neurodegeneration? Autophagy *2*, 302-304.

Mizushima, N., and Klionsky, D.J. (2007). Protein turnover via autophagy: implications for metabolism. Annual review of nutrition *27*, 19-40.

Mizushima, N., Levine, B., Cuervo, A.M., and Klionsky, D.J. (2008). Autophagy fights disease through cellular self-digestion. Nature *451*, 1069-1075.

Mizushima, N., Yamamoto, A., Matsui, M., Yoshimori, T., and Ohsumi, Y. (2004). *In vivo* analysis of autophagy in response to nutrient starvation using transgenic mice expressing a fluorescent autophagosome marker. Molecular biology of the cell *15*, 1101-1111.

Mizushima, N., and Yoshimori, T. (2007). How to interpret LC3 immunoblotting. Autophagy *3*, 542-545.

Mok, S.W., Riemer, C., Madela, K., Hsu, D.K., Liu, F.T., Gultner, S., Heise, I., and Baier, M. (2007). Role of galectin-3 in prion infections of the CNS. Biochemical and biophysical research communications *359*, 672-678.

Moore, R.C., Lee, I.Y., Silverman, G.L., Harrison, P.M., Strome, R., Heinrich, C., Karunaratne, A., Pasternak, S.H., Chishti, M.A., Liang, Y., *et al.* (1999). Ataxia in prion protein (PrP)-deficient mice is associated with upregulation of the novel PrP-like protein doppel. Journal of molecular biology *292*, 797-817.

Nakai, A., Yamaguchi, O., Takeda, T., Higuchi, Y., Hikoso, S., Taniike, M., Omiya, S., Mizote, I., Matsumura, Y., Asahi, M., *et al.* (2007). The role of autophagy in cardiomyocytes in the basal state and in response to hemodynamic stress. Nature medicine *13*, 619-624.

Nixon, R.A., Wegiel, J., Kumar, A., Yu, W.H., Peterhoff, C., Cataldo, A., and Cuervo, A.M. (2005). Extensive involvement of autophagy in Alzheimer disease: an immuno-electron microscopy study. Journal of neuropathology and experimental neurology *64*, 113-122.

Oh, J.M., Shin, H.Y., Park, S.J., Kim, B.H., Choi, J.K., Choi, E.K., Carp, R.I., and Kim, Y.S. (2008). The involvement of cellular prion protein in the autophagy pathway in neuronal cells. Molecular and cellular neurosciences *39*, 238-247.

Pankiv, S., Clausen, T.H., Lamark, T., Brech, A., Bruun, J.A., Outzen, H., Overvatn, A., Bjorkoy, G., and Johansen, T. (2007). p62/SQSTM1 binds directly to Atg8/LC3 to facilitate degradation of ubiquitinated protein aggregates by autophagy. The Journal of biological chemistry *282*, 24131-24145.

Parkin, E.T., Watt, N.T., Turner, A.J., and Hooper, N.M. (2004). Dual mechanisms for shedding of the cellular prion protein. The Journal of biological chemistry *279*, 11170-11178.

Petersen, A., Larsen, K.E., Behr, G.G., Romero, N., Przedborski, S., Brundin, P., and Sulzer, D. (2001). Expanded CAG repeats in exon 1 of the Huntington's disease gene stimulate dopamine-mediated striatal neuron autophagy and degeneration. Human molecular genetics *10*, 1243-1254.

Prusiner, S.B. (1982). Novel proteinaceous infectious particles cause scrapie. Science (New York, NY *216*, 136-144.

Prusiner, S.B. (1998). Prions. Proceedings of the National Academy of Sciences of the United States of America *95*, 13363-13383.

Qin, Z.H., Wang, Y., Kegel, K.B., Kazantsev, A., Apostol, B.L., Thompson, L.M., Yoder, J., Aronin, N., and DiFiglia, M. (2003). Autophagy regulates the processing of amino terminal huntingtin fragments. Human molecular genetics *12*, 3231-3244.

Qu, X., Yu, J., Bhagat, G., Furuya, N., Hibshoosh, H., Troxel, A., Rosen, J., Eskelinen, E.L., Mizushima, N., Ohsumi, Y., *et al.* (2003). Promotion of tumorigenesis by heterozygous disruption of the beclin 1 autophagy gene. The Journal of clinical investigation *112*, 1809-1820.

Ravikumar, B., Duden, R., and Rubinsztein, D.C. (2002). Aggregate-prone proteins with polyglutamine and polyalanine expansions are degraded by autophagy. Human molecular genetics *11*, 1107-1117.

Ravikumar, B., Vacher, C., Berger, Z., Davies, J.E., Luo, S., Oroz, L.G., Scaravilli, F., Easton, D.F., Duden, R., O'Kane, C.J., *et al.* (2004). Inhibition of mTOR induces autophagy and reduces toxicity of polyglutamine expansions in fly and mouse models of Huntington disease. Nature genetics *36*, 585-595.

Reggiori, F., and Klionsky, D.J. (2002). Autophagy in the eukaryotic cell. Eukaryotic cell *1*, 11-21.

Rossi, D., Cozzio, A., Flechsig, E., Klein, M.A., Rulicke, T., Aguzzi, A., and Weissmann, C. (2001). Onset of ataxia and Purkinje cell loss in PrP null mice inversely correlated with Dpl level in brain. The EMBO journal *20*, 694-702.

Rubinsztein, D.C. (2006). The roles of intracellular protein-degradation pathways in neurodegeneration. Nature *443*, 780-786.

Rubinsztein, D.C., DiFiglia, M., Heintz, N., Nixon, R.A., Qin, Z.H., Ravikumar, B., Stefanis, L., and Tolkovsky, A. (2005). Autophagy and its possible roles in nervous system diseases, damage and repair. Autophagy *1*, 11-22.

Rubinsztein, D.C., Gestwicki, J.E., Murphy, L.O., and Klionsky, D.J. (2007). Potential therapeutic applications of autophagy. Nat Rev Drug Discov *6*, 304-312.

Sakaguchi, S., Katamine, S., Nishida, N., Moriuchi, R., Shigematsu, K., Sugimoto, T., Nakatani, A., Kataoka, Y., Houtani, T., Shirabe, S., *et al.* (1996). Loss of cerebellar Purkinje cells in aged mice homozygous for a disrupted PrP gene. Nature *380*, 528-531.

Sarkar, S., Davies, J.E., Huang, Z., Tunnacliffe, A., and Rubinsztein, D.C. (2007a). Trehalose, a novel mTOR-independent autophagy enhancer, accelerates the clearance of mutant huntingtin and alpha-synuclein. The Journal of biological chemistry *282*, 5641-5652.

Sarkar, S., Floto, R.A., Berger, Z., Imarisio, S., Cordenier, A., Pasco, M., Cook, L.J., and Rubinsztein, D.C. (2005). Lithium induces autophagy by inhibiting inositol monophosphatase. The Journal of cell biology *170*, 1101-1111.

Sarkar, S., Perlstein, E.O., Imarisio, S., Pineau, S., Cordenier, A., Maglathlin, R.L., Webster, J.A., Lewis, T.A., O'Kane, C.J., Schreiber, S.L., *et al.* (2007b). Small molecules enhance autophagy and reduce toxicity in Huntington's disease models. Nature chemical biology *3*, 331-338.

Schatzl, H.M., Laszlo, L., Holtzman, D.M., Tatzelt, J., DeArmond, S.J., Weiner, R.I., Mobley, W.C., and Prusiner, S.B. (1997). A hypothalamic neuronal cell line persistently infected with scrapie prions exhibits apoptosis. Journal of virology *71*, 8821-8831.

Seglen, P.O., and Gordon, P.B. (1982). 3-Methyladenine: specific inhibitor of autophagic/lysosomal protein degradation in isolated rat hepatocytes. Proceedings of the National Academy of Sciences of the United States of America *79*, 1889-1892.

Shibata, M., Lu, T., Furuya, T., Degterev, A., Mizushima, N., Yoshimori, T., MacDonald, M., Yankner, B., and Yuan, J. (2006). Regulation of intracellular accumulation of mutant Huntingtin by Beclin 1. The Journal of biological chemistry *281*, 14474-14485.

Sikorska, B., Liberski, P.P., Giraud, P., Kopp, N., and Brown, P. (2004). Autophagy is a part of ultrastructural synaptic pathology in Creutzfeldt-Jakob disease: a brain biopsy study. The international journal of biochemistry & cell biology *36*, 2563-2573.

Silveira, J.R., Raymond, G.J., Hughson, A.G., Race, R.E., Sim, V.L., Hayes, S.F., and Caughey, B. (2005). The most infectious prion protein particles. Nature *437*, 257-261.

Sunyach, C., Jen, A., Deng, J., Fitzgerald, K.T., Frobert, Y., Grassi, J., McCaffrey, M.W., and Morris, R. (2003). The mechanism of internalization of glycosylphosphatidylinositol-anchored prion protein. The EMBO journal 22, 3591-3601.

Suzuki, K., and Ohsumi, Y. (2007). Molecular machinery of autophagosome formation in yeast, Saccharomyces cerevisiae. FEBS letters *581*, 2156-2161.

Takahashi, Y., Coppola, D., Matsushita, N., Cualing, H.D., Sun, M., Sato, Y., Liang, C., Jung, J.U., Cheng, J.Q., Mule, J.J., *et al.* (2007). Bif-1 interacts with Beclin 1 through UVRAG and regulates autophagy and tumorigenesis. Nature cell biology *9*, 1142-1151.

Taraboulos, A., Scott, M., Semenov, A., Avrahami, D., Laszlo, L., and Prusiner, S.B. (1995). Cholesterol depletion and modification of COOH-terminal targeting sequence of the prion protein inhibit formation of the scrapie isoform. The Journal of cell biology *129*, 121-132.

Tsukada, M., and Ohsumi, Y. (1993). Isolation and characterization of autophagy-defective mutants of Saccharomyces cerevisiae. FEBS letters *333*, 169-174.

Ventruti, A., and Cuervo, A.M. (2007). Autophagy and neurodegeneration. Current neurology and neuroscience reports *7*, 443-451.

Webb, J.L., Ravikumar, B., Atkins, J., Skepper, J.N., and Rubinsztein, D.C. (2003). Alpha-Synuclein is degraded by both autophagy and the proteasome. The Journal of biological chemistry *278*, 25009-25013.

Yorimitsu, T., and Klionsky, D.J. (2007). Eating the endoplasmic reticulum: quality control by autophagy. Trends in cell biology *17*, 279-285.

Yoshimori, T. (2004). Autophagy: a regulated bulk degradation process inside cells. Biochemical and biophysical research communications *313*, 453-458.

Yu, W.H., Cuervo, A.M., Kumar, A., Peterhoff, C.M., Schmidt, S.D., Lee, J.H., Mohan, P.S., Mercken, M., Farmery, M.R., Tjernberg, L.O., *et al.* (2005). Macroautophagy--a novel Beta-amyloid peptide-generating pathway activated in Alzheimer's disease. The Journal of cell biology *171*, 87-98.

Yue, Z., Jin, S., Yang, C., Levine, A.J., and Heintz, N. (2003). Beclin 1, an autophagy gene essential for early embryonic development, is a haploinsufficient tumor suppressor. Proceedings of the National Academy of Sciences of the United States of America *100*, 15077-15082.

Yun, S.W., Ertmer, A., Flechsig, E., Gilch, S., Riederer, P., Gerlach, M., Schatzl, H.M., and Klein, M.A. (2007). The tyrosine kinase inhibitor imatinib mesylate delays prion neuroinvasion by inhibiting prion propagation in the periphery. J Neurovirol *13*, 328-337.

Chapter 4

Prion Protein and Metal Interaction: Physiological and Pathological Implications

Neena Singh, Dola Das, Ajay Singh and Maradumane L. Mohan

Abstract

Metal induced free radicals are important mediators of neurotoxicity in several neurodegenerative conditions such as Alzheimer's disease, Parkinson's disease, and Huntington's disease. Similar evidence is now emerging for prion diseases, a group of neurodegenerative disorders of humans and animals. The main pathogenic agent in all prion disorders is PrP-scrapie (PrP^{Sc}), a β-sheet rich isoform of a normal cell surface glycoprotein known as the prion protein (PrP^{C}). Deposits of PrP^{Sc} in the brain parenchyma are believed to induce neurotoxicity through poorly understood mechanisms. Recent reports suggest that imbalance of brain metal homeostasis is a significant cause of PrP^{Sc}-associated neurotoxicity, though the underlying mechanisms are difficult to explain based on existing information. Proposed hypotheses include a functional role for PrP^{C} in metal metabolism, and loss of this function due to aggregation to the disease associated PrP^{Sc} form as the cause of brain metal imbalance. Other views suggest gain of toxic function by PrP^{Sc} due to sequestration of PrP^{C}-associated metals within the aggregates, resulting in the generation of redox-active PrP^{Sc} complexes. The physiological implications of some PrP^{C}-metal interactions are known, while others are still unclear. The pathological implications of PrP^{C}-metal interaction include metal-induced oxidative damage, and in some instances conversion of PrP^{C} to a PrP^{Sc}-like form. Despite its significance, only limited information is available on PrP-metal interaction and its implications on prion disease pathogenesis. In this review, we summarize the physiological significance and pathological implications of PrP-metal interaction on prion disease pathogenesis.

Abbreviations

Fe^{2+}: Ferrous iron
Fe^{3+}: Ferric iron
AD: Alzheimer's disease
CJD: Creutzfeldt-Jakob disease
HFE: hemochromatosis protein
HD: Huntington's disease
LIP: labile iron pool
N2a: mouse neuroblastoma cells
PD: Parkinson's disease
PK: proteinase-K
$PrP^{-/-}$: mice lacking PrP expression
PrP^{C}: prion protein
PrP^{Sc}: PrP-scrapie
ScN2a: scrapie infected mouse neuroblastoma cells
SOD: Cu/Zn superoxide dismutase
Tf: transferrin, TfR: transferrin receptor

Introduction

The current challenge facing prion research is the lack of an effective therapeutic strategy for prion disorders, a group of invariably fatal neurodegenerative conditions of humans and animals. Research in this area has been hampered due to incomplete understanding of the underlying mechanisms of neurotoxicity in these devastating diseases. The most favored hypothesis supported by numerous studies suggests that neurotoxicity in all prion disorders is mediated by PrP-scrapie (PrP^{Sc}), a β-sheet rich conformation of a cell surface glycoprotein known as the prion protein (PrP^{C}). The change in conformation of PrP^{C} to PrP^{Sc} confers certain biochemical and biophysical properties to PrP^{Sc}, which, unlike its counterpart PrP^{C}, becomes insoluble in non-ionic detergents and resists limited digestion by proteinase-K (PK). Deposits of PrP^{Sc} in the brain parenchyma are considered the principal cause of neurotoxicity, though the pathways involved in this process are poorly understood (Prusiner, 1998; Aguzzi and Polymenidou, 2004, Caughey and Baron, 2006). Intense research in this area has clarified several aspects of this process. It is now clear that PrP^{Sc} in the extracellular space does not induce toxicity in the absence of PrP^{C} expression on the neuronal plasma membrane, implicating PrP^{C} in mediating the toxic signal (Malluci et al., 2003; Chesebro et al., 2005). On the other hand, accumulation of PrP^{Sc} only on astrocytes induces neurotoxicity, suggesting indirect pathways of toxicity by PrP^{Sc} (Jeffrey et al., 2004). Suggested pathways include secretion of toxic chemokines and factors by microglia in response to PrP^{Sc}, physical damage to the membrane structure by PrP^{Sc} aggregates, interference with synaptic transmission, and toxicity through a labile by-product of the PrP^{C} to PrP^{Sc} conversion reaction (reviewed in Harris and True, 2006; Caughey and Baron, 2006). Evidence from cell model studies of familial prion disorders indicates abnormal processing and metabolism of mutant PrP as a possible cause of cytotoxicity, though the mechanism varies with specific mutations (Jin et al., 2000; Mishra et al., 2002; Gu et al., 2003). Other mechanisms such as activation of cell death pathways due to intracellular accumulation of PrP^{Sc} (Yadavalli et al., 2004; Kristiansen et al., 2005), cross-linking of neuronal PrP^{C} on the plasma membrane by PrP^{Sc} (Solforosi et al., 2004), and toxicity due to C-transmembrane

The Prion Protein 2010 Jörg Tatzelt (Ed.) Savanna Press, UK
ISBN: 978-0-9543335-2-2. savannapress.com

and cytosolic forms of PrP^C through unspecified pathways have also been suggested, but the details of underlying mechanisms remain unclear (Hegde et al., 1998; 1999; Ma et al., 2002; Ma and Lindquist, 2002; Heller et al., 2003; Wang et al., 2006).

An overlooked but equally important cause of neurotoxicity is the loss of normal function of PrP^C due to conversion to the PrP^{Sc} form. Information on this subject is limited since the normal function of PrP^C remains ambiguous despite its ubiquitous presence and conservation through evolution. Transgenic mice lacking PrP^C expression ($PrP^{-/-}$) do not develop a specific phenotype unless a prion-like protein Doppel is up regulated, making it difficult to ascertain the functional role of PrP^C (Bueler et al., 1992). Other experimental models suggest several unrelated functions, including a role in signal transduction, cell adhesion, copper uptake and transport, as an anti-oxidant and anti-apoptotic protein, and others (Bounhar et al., 2001; Roucou et al., 2004; 2005; Roucou and LeBlanc, 2005; Harris and True, 2006). The diverse functions attributed to PrP^C probably reflect its involvement in an essential biochemical pathway that affects several cellular processes. Loss of this activity by the change in conformation of PrP^C to PrP^{Sc} is likely to induce toxicity due to loss of an essential function, together with gain of certain toxic properties by PrP^{Sc}.

One such pathway may involve the functional role of PrP^C in copper and iron metabolism. Since these metals are redox-active, abnormal metabolism of either of these metals is likely to induce neurotoxicity due to the generation of free radicals. Such a mechanism has been observed in neurodegenerative conditions associated with protein aggregation such as Alzheimer's disease (AD) and Parkinson's disease (PD) (Moos and Morgan, 2004; Gerlach et al., 1994; Hayashi et al., 2006; Gaasch et al., 2007; Smith et al., 1997; 2007; Barnham and Bush, 2008). Recent reports indicate the presence of a similar phenomenon in prion disorders (Petersen et al., 2005; Singh et al., 2009a). Markers of oxidative stress and imbalance of metal homeostasis have been reported in prion disease affected brains, lending support to this assumption (Kim et al., 2000; Rossi et al 2004; Petersen et al., 2005; Pamplona et al., 2008; Singh et al., 2009a). In this review, we summarize evidence related to the interaction of PrP^C with various metals, followed by the physiological and pathological implications of PrP-metal interaction. In particular, the role of PrP^C in copper and iron metabolism is reviewed since these metals are essential for several metabolic processes, and are also toxic if mismanaged due to their redox-active nature.

PrP-metal interactions

Several studies have reported the interaction of PrP^C with metals using *in vitro* and *in vivo* models. The *in vitro* studies have been more revealing due to the simplicity and accuracy of the readout compared to cell and animal models where metal metabolism is complex and the interaction of individual proteins with metals is often missed due to low affinity or their transient nature. Observations from *in vitro* studies using re-folded, full-length recombinant PrP or its fragments have lead to important findings regarding PrP-metal interaction

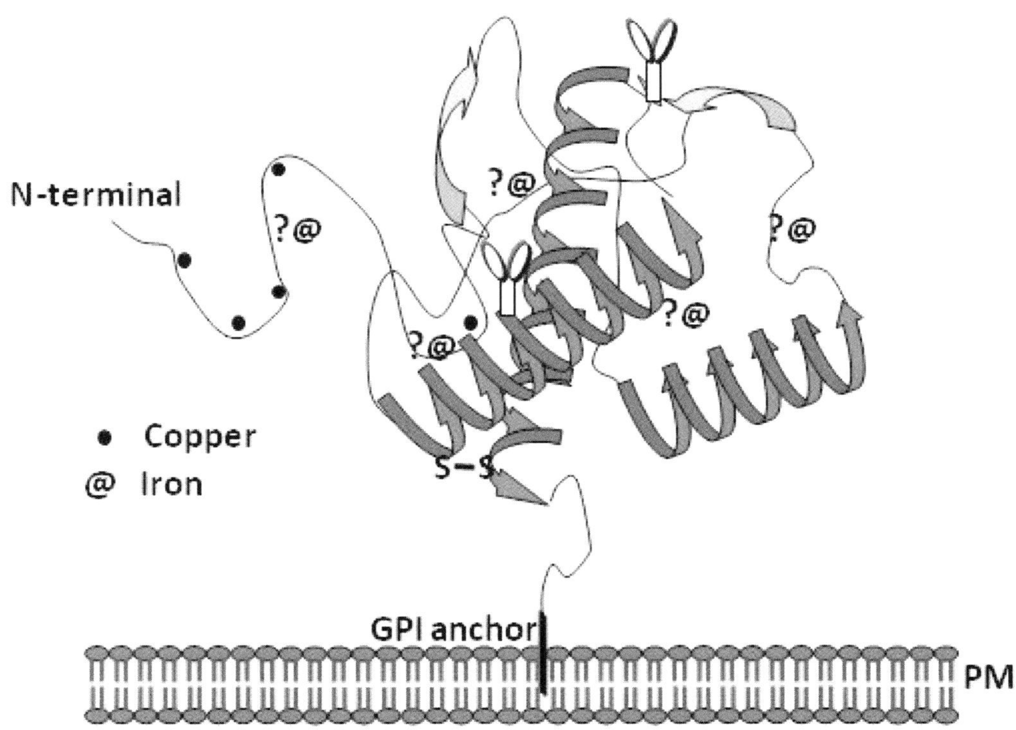

Figure 1. A model of PrP^C demonstrating the known copper binding sites in the octa-peptide region and histidine residues 96 and 111. The association of PrP^C with iron is based on unpublished data from recombinant full length PrP^C and its fragments.

and its significance to prion disease pathogenesis. It is now clear that recombinant PrP binds several divalent cations, including copper, iron, zinc, manganese and nickel (Pan et al., 1992; Hornshaw et al., 1995a; 1995b; Brown et al., 2000; Jackson et al., 2001; Jones et al., 2004; 2005; Basu et al., 2007). The highly conserved octa-peptide repeat region of PrP is the principal metal binding site, and its affinity for metals is highest for copper, followed by nickel, zinc, and manganese (Jackson et al. 2001). A diagrammatic representation of the interaction of PrP^C with copper and iron is shown in Figure 1. The interaction of PrP^C with metals is important to understand because of the physiological and pathological implications of this association. For example, PrP^C is believed to mediate the uptake of copper and iron, suggesting a role in cellular copper and iron metabolism (Brown and Harris, 2003; Singh et al., 2009b,c). On the other hand, interaction of recombinant and cell associated PrP^C with certain metals induces a change in its conformation to the disease associated PrP^{Sc} form, suggesting their role in prion disease pathogenesis (Deleaut et al., 2007; Basu et al., 2007). The conversion of PrP^C to the PrP^{Sc} form increases its affinity for nickel while decreasing the binding of zinc and manganese, indicating specific interaction of different metals with the normal and the disease associated PrP^{Sc} form (Jackson et al., 2001).

The association of PrP^C with copper is better characterized than its interaction with other metals. Four copper binding sites have been identified within the octa-peptide repeat region of PrP^C with additional sites on histidine residues 96 and 111 (Burns et al., 2003). Purified PrP^C from mouse and human brains also binds copper, substantiating the *in vitro* observations (Brown et al., 2001; Wong et al., 2001b). The affinity of PrP^C for copper is higher compared to zinc, and even a large excess of zinc cannot displace copper from the octa-peptide repeat region of full-length PrP^C under *in vitro* conditions (Jackson et al., 2001; Walter et al., 2007). However, physiologically relevant levels of zinc in cell models alter the distribution of PrP^C bound copper with relative ease, indicating the influence of other catalytic factors in determining the metal ion occupancy of PrP^C (Watt and Hooper, 2003). These observations suggest that the interaction of PrP^C with zinc may be more significant given the relative abundance of this metal in the brain (Qin et al., 2002; Watt and Hooper, 2003; Walter at al., 2007; Kenward et al., 2007). PrP^C also binds manganese, probably in the C-terminal region between residues 91-230 or overlapping with the copper binding site at His-96 (Treiber et al., 2007; Brazier et al., 2008). Although these observations are useful, the data need to be interpreted with caution since *in vitro* reactions do not always represent the complex milieu of PrP^C in cells or in the brain.

The interaction of PrP^C with some of these metals alters its structure such that it simulates PrP^{Sc} in certain biochemical properties including detergent insolubility and resistance to limited digestion by proteinase-K. Other metals induce this change in PrP^C by their absence such as in the presence of a specific chelator, suggesting that modification of metal ion occupancy of PrP^C is equally significant in bringing about this change (Deleaut et al., 2007). Thus, addition of manganese promotes a change in the conformation of recombinant PrP^C to PrP^{Sc}, while zinc is believed to inhibit fibril formation and promote inter-molecular reactions (Brown et al., 2000; Giese et al., 2004; Tsenkova et al., 2004; Bocharova et al., 2005; Kim et al., 2005; Abdelraheim et al, 2006; Treiber et al., 2006; Kenward et al., 2007). Addition of zinc and copper to the toxic peptide of PrP, a model often used for studying prion-mediated toxicity, increases its aggregation and toxicity to cells (Jobling et al., 1999; 2001). A similar change in the conformation of PrP^C to a PrP^{Sc}-like form is noted when PrP-expressing yeast cells are grown in medium supplemented with copper or manganese (Treiber et al., 2006). This reaction is reversed by adding specific chelators such as bathocuproinedisulfonic acid and clioquinol to the medium, reinforcing the idea that only specific metals induce this reaction. Despite this information and additional data emphasizing the significance of PrP^C and PrP^{Sc}-metal interaction, neither the physiological nor the pathological significance of this interaction is clearly understood.

Physiological significance of PrP-metal interaction

Observations from neuroblastoma cells suggest that PrP^C binds extracellular copper ions and delivers to endocytic compartments, functioning as a copper uptake protein (Pauly and Harris, 1998; Brown and Harris, 2003). The octa-peptide repeats of PrP^C are essential for this process since deletion of this region inhibits copper uptake (Perera and Hooper, 2001). The octa-peptide repeat region is also believed to function as a reductase, reducing Cu (II) ions before transport across the endosomal membrane to the cytosol (Miura et al., 2005). Although these observations suggest that PrP^C may be a major copper delivery protein, surprisingly, there is minimal difference in the copper content of brains from wild type and transgenic mice lacking PrP expression ($PrP^{-/-}$) (Giese et al., 2005). However, contradictory studies claim a significant difference in copper levels between wild type, $PrP^{-/-}$, and scrapie infected mouse brains, suggesting a prominent role for PrP^C in maintaining copper homeostasis in the brain (Brown et al., 1997, 1998; Wong et al., 2001c; Thackray et al., 2002). Likewise, conflicting results regarding brain copper levels have been reported in PrP^C over-expressing mice (Herms et al., 1999; Kretzschmar et al., 2000; Waggoner et al., 2000; Stuermer and Plattner, 2005), leaving the matter unsettled.

The interaction of PrP^C with iron is a relatively recent finding, and the physiological significance of this association is becoming increasingly clear from studies indicating a functional role for PrP^C in cellular iron uptake and transport (Basu et al., 2007; Singh et al., 2009b; 2009c). Unlike most other divalent cations, iron is an essential component of enzymes and proteins and is required for optimal neuronal growth and function (Beard and Connor, 2003). On the other hand, iron is also considered a toxin due to its ability to exist in two oxidation states (ferric Fe^{3+} and ferrous Fe^{2+}) (Thompson et al., 2001; Kaplan, 2002). Due to this reason, iron transport into and out of the cells is tightly regulated. It is surprising that PrP^C influences the cellular iron pool within this tightly regulated mechanism of iron uptake, transport, and utilization (Singh et al., 2009b; 2009c). These observations have significant bearing on the pathogenesis of prion disorders since aggregation of PrP^C is likely to disturb cellular iron homeostasis, resulting in neurotoxicity.

PrP^C has been demonstrated to influence iron metabolism in cells expressing normal and mutant PrP forms and in $PrP^{-/-}$ transgenic mouse models (Singh et al., 2009b, 2009c). Over-expression of PrP^C in cultured neuroblastoma cells increases the cellular labile iron pool

(LIP) and iron saturation of ferritin, suggesting a role for PrP^C in iron uptake. It is interesting to note that cells expressing pathogenic and non-pathogenic mutations of PrP alter the cellular LIP and iron saturation of ferritin differentially. This difference in the cellular iron content is maintained even when cells are cultured in the presence of excess extracellular iron, indicating a dominant role in iron uptake (Singh et al., 2009b). Furthermore, stimulation of endocytosis by a PrP-specific antibody increases intracellular iron stores, suggesting that PrP^C mediates uptake of iron from the extracellular milieu. Unlike ceruloplasmin, PrP^C does not mediate the efflux of excess iron from cells, confirming its role as an iron uptake protein (Jeong and David, 2003; Singh et al., 2009b). Presently it is unclear whether PrP^C mediates iron uptake using a novel pathway or by interacting with the conventional pathway of iron uptake and transport. It has been hypothesized that PrP^C may influence iron uptake by interacting with the transferrin/transferrin receptor pathway as described for HFE (Waheed et al., 2002), or function as a ferric reductase to facilitate the transport of ferric iron from endosomes to cytosolic ferritin (Singh et al., 2009b).

Similar observations suggesting a positive effect of PrP^C on systemic iron levels are noted when wild type mice are compared with $PrP^{-/-}$ mouse models. Deletion of PrP^C in $PrP^{-/-}$ mice induces iron deficiency in the latter by decreasing the efficiency of iron transport from the intestinal lumen to the blood stream, and uptake of iron from the blood by parenchymal cells and cells of the hematopoietic lineage. Re-expression of PrP^C corrects the iron deficiency in these mice, confirming the functional role for PrP in iron uptake (Singh et al., 2009c). Considering that $PrP^{-/-}$ mice are only mildly iron deficient, it is likely that PrP^C modulates the function of other iron uptake proteins or is involved in a pathway that compensates for its absence.

An important question that remains unanswered is the binding site and the affinity of PrP^C for iron. *In vitro* experiments using recombinant PrP^C and its fragments indicate that the octapeptide repeat region of PrP^C is not essential for iron binding, and the iron and copper binding regions of PrP^C do not overlap (unpublished observations). Limited data using different denaturing conditions suggest that the interaction of PrP^C with iron depends on the

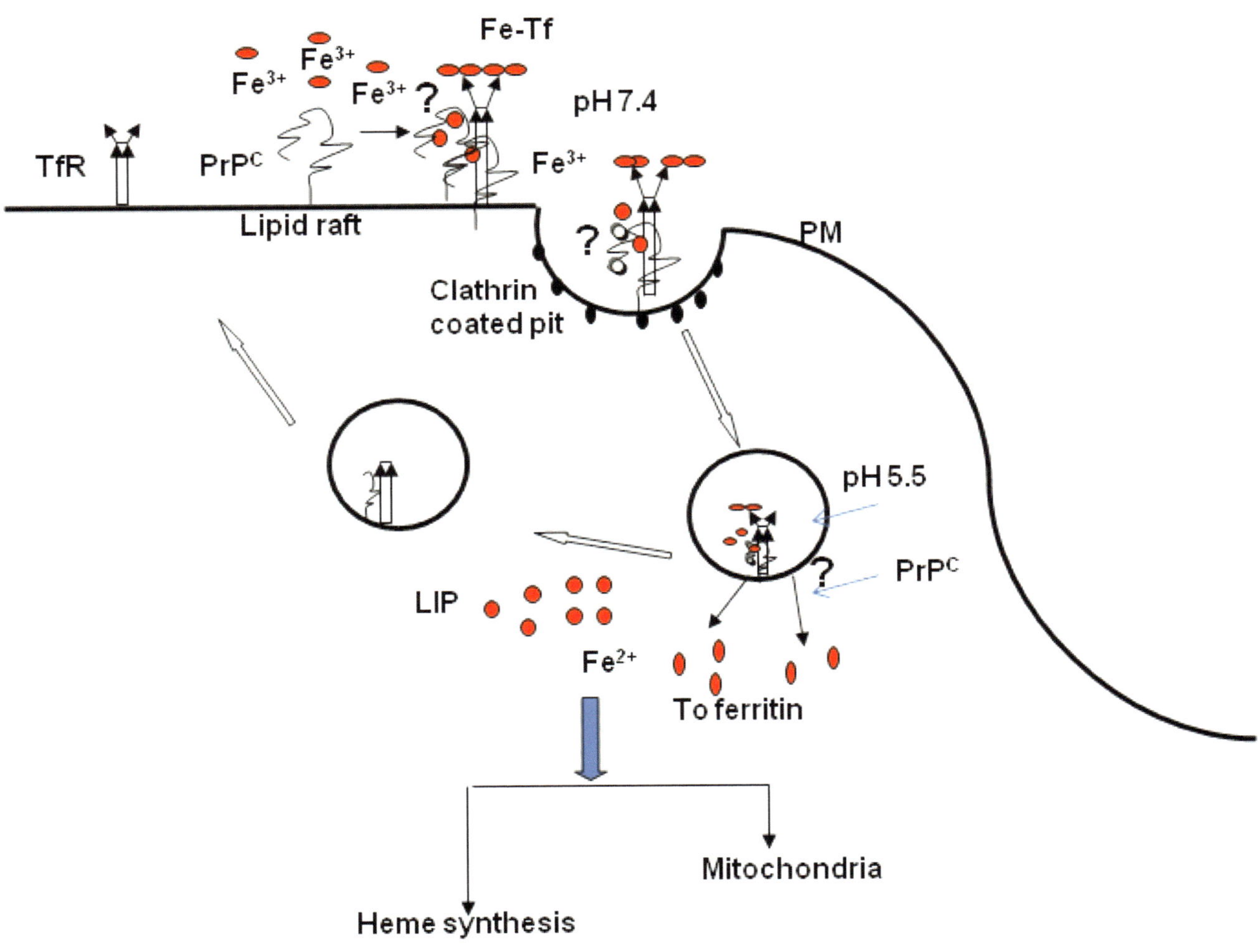

Figure 2. A representation of possible mechanisms by which PrP^C is likely to mediate cellular iron uptake. At the plasma membrane, PrP^C may bind iron directly for transport to the cytosol through endocytosis. It is also likely that PrP^C influences iron uptake indirectly by modulating the endocytosis association or endocytosis of the Tf/TfR complex. Alternately, PrP^C may facilitate the transport of iron from the endosomal compartment to the cytosol by functioning as a ferric-reductase. In the cytosol, iron is used for metabolic processes, or stored in ferritin in a relatively inert form.

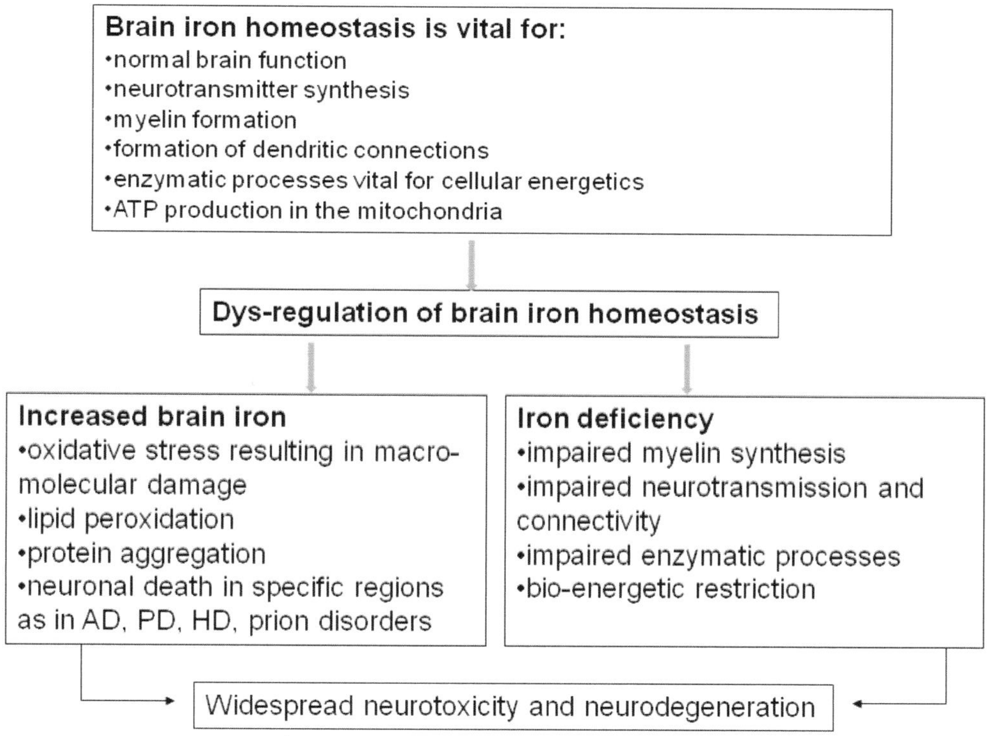

Figure 3. A model demonstrating the vital role of iron in brain function such as neurotransmitter synthesis, myelin formation, maintenance of dendritic connections, and other metabolic processes. Dys-regulation of brain iron homeostasis can induce neurotoxicity both by iron excess, and by iron deficiency. The former is a consequence of the redox-active nature of iron leading to oxidative stress, and the latter due to impairment of essential metabolic processes.

conformation of PrP^C rather than a specific amino acid sequence. These studies are limited to recombinant PrP^C since it is difficult to identify iron associated cellular PrP^C, suggesting that either the interaction of PrP^C with iron is transient, or the affinity of PrP^C for iron is not high enough for easy identification. A diagrammatic representation of possible mechanisms of iron uptake by PrP^C is depicted in Figure 2. It is likely that after binding iron, PrP^C moves from lipid rich domains in the plasma membrane to the vicinity of TfR, where it might influence the binding of Tf to the TfR, internalization of PrP^C/TfR/Tf complex, or endocytosis of Tf/TfR complex. It is also possible that PrP^C influences iron transport from the endosomes to the cytosol by functioning as a ferric-reductase as demonstrated for copper (Miura et al., 2005).

Pathological implications of PrP-metal interaction

Copper and iron have significant pathological implications due to their redox-active nature. For example, both copper and iron induce the conversion of PrP^C to the PrP^{Sc} form (Kim et al., 2005). Addition of copper increases the protease resistance and infectivity of denatured PrP^{Sc} *in vitro*, while copper chelation *in vivo* delays the onset of disease in prion infected mice (Pauly and Harris, 1998; Quaglio et al., 2001; Sigurdsson et al., 2003; Kuczius et al., 2004). Paradoxically, copper treatment inhibits PrP^{Sc} formation in cells infected with mouse prions and delays the onset of disease in scrapie infected hamsters, demonstrating a protective role, perhaps by augmenting the function of PrP^C as a Cu/Zn superoxide dismutase (SOD) (Vassallo and Herms, 2003; Hijazi et al., 2003; Kiachopoulos et al., 2004; Orem et al., 2006). These conflicting observations can perhaps be explained from *in vitro* studies where copper inhibits the amplification of PrP^{Sc} from purified brain-derived PrP^C and recombinant PrP^C by stabilizing its α-helical structure, while it enhances the β-sheet structure of preformed PrP fibrils, thereby increasing their PrP^{Sc} content (Liu et al., 2007). Copper could therefore delay or augment disease progression based on the time when it is introduced to the animal.

An important consideration in evaluating the pathogenic consequences of PrP^C-metal interaction is the cellular compartment in which metal bound PrP^C is exposed to free radicals. It has been observed that exposure of PrP^C expressing cells to a source of redox-active metal such as ferrous chloride induces the generation of PrP^{Sc}-like aggregates on the plasma membrane that accumulate within lysosomes in association with ferritin (Basu et al., 2007). These aggregates are also redox-active and initiate the generation of additional PrP-ferritin aggregates, propagating the PrP^{Sc}-like conformation within cells (Basu et al., 2007). Likewise, exposure of PrP^C expressing cells to hemin, an iron containing compound, also results in the aggregation and degradation of PrP^C (Lee et al., 2007), implicating redox-iron in the generation and propagation of PrP^{Sc}. Similar aggregation of α-synuclein is noted in response to iron and copper (Paik et al., 2000; Ostrerova-Golts et al., 2000; Golts et al., 2002; Takahashi et al., 2007), suggesting that protein aggregation by redox-active metals is not specific to PrP^C. However, the generation of additional PrP^{Sc} aggregates by redox-active PrP^{Sc} appears to be unique to PrP, explaining in part the propagation of PrP^{Sc} once initiated. Surprisingly, chelation of iron from diseased brain homogenates decreases the total amount of PK-resistant PrP^{Sc}, suggesting that iron is also involved in the stability of PrP^{Sc} (Basu et al., 2007). This observation is of immense prophylactic value since iron chelation can be used as a means to decrease the infectivity of prion contaminated material.

Observations from mouse models of prion disease demonstrate an increase in the levels of iron regulatory proteins 1 and 2 and iron storage protein ferritin in the hippocampus and cerebral cortex of diseased brains, indicating the presence of brain iron imbalance (Kim et al., 2007). A similar imbalance of cellular iron homeostasis is observed in scrapie infected mouse neuroblastoma cells (ScN2a) that also show increased susceptibility to iron induced oxidative stress (Fernaeus et al., 2005; Fernaeus and Land, 2005). Imbalance of iron homeostasis is also observed in prion disease affected human, mouse, and hamster brains that demonstrate a phenotype of iron deficiency in the presence of increased total and redox-active Fe^{2+} iron (Singh et al., 2009a). The iron deficiency in scrapie infected hamster brains is noted ~6 weeks after inoculation of the infectious material and correlates with PrP^{Sc} levels, perhaps due to sequestration of iron in PrP^{Sc}-ferritin complexes (Singh et al., 2009a). The redox-active nature of PrP^{Sc}-ferritin aggregates is likely to induce the aggregation of additional PrP^{C}, creating an ongoing state of iron imbalance and associated neurotoxicity in the diseased brain (Mishra et al., 2004; Basu et al., 2007). The associated iron deficiency caused by PrP^{Sc}-ferritin aggregates is equally harmful especially since iron is required for several vital metabolic processes (Figure 3).

In addition to copper and iron, alteration in the homeostasis of manganese has also been reported in prion disease affected human and animal brains. In scrapie infected mice, levels of manganese are elevated in the peripheral blood, while in diseased cattle and sheep increased levels are noted in both the blood and brain tissue (Wong et al., 2001a, 2001b; Thackray et al., 2002; Hesketh et al. 2007; 2008). Surprisingly, altered levels of manganese have also been reported in scrapie resistant sheep that do not develop clinical disease, making it unlikely that the change is due to the disease process *per se*. However, limited studies make it difficult to explain the generality or the underlying cause of this abnormality. Other diseases that show elevated levels of manganese include hemochromatosis and diseases of the blood-brain barrier, information that may help in understanding the basis of manganese imbalance in prion disorders. Other reports indicate a decrease in the levels of copper and zinc in the brains of scrapie infected mice, although blood levels of zinc increase during disease progression as noted for manganese (Wong et al. 2001a, 2001b; Thackray et al., 2002). In human brains affected with sCJD, PrP^{C}-associated copper is replaced by zinc, resulting in the loss of SOD-like activity of PrP^{C}, a possible cause of increased neurotoxicity due to oxidative stress. Further studies are necessary to understand this phenomenon fully.

Summary and Perspective

Despite overwhelming evidence implicating PrP^{Sc} in the pathogenesis of prion disorders, the mechanistic details underlying the neurotoxicity associated with these disorders remain unclear. The association of PrP^{C} with redox-active metals such as copper and iron provides new insight into the role of metal-induced oxidative stress in these disorders. Observations from cell and mouse models suggest that PrP^{C} is involved in copper and iron uptake. Although PrP^{C} is not a major iron or copper modulating protein, compromised levels of these metals due to aggregation of PrP^{C} to the PrP^{Sc} form are likely to affect cell health due to their redox-active nature. At the same time, PrP^{Sc} assumes a redox-active nature due to sequestration of iron in PrP^{Sc}-ferritin aggregates, providing a logical explanation for imbalance of iron homeostasis in prion disease affected brains. This phenotype is likely to induce oxidative stress and neuronal damage, and could contribute significantly to prion disease associated neurotoxicity. Although the role of redox-active metals in prion disease pathogenesis is still at its formative stage, sufficient information exists to promt future investigations on this subject that may lead to the development of anti-oxidants and metal chelators as useful therapeutic agents.

References

Aguzzi, A., and Polymenidou, M. (2004). Mammalian prion biology: one century of evolving concepts. Cell. *116*, 313-327.

Abdelraheim, S.R., Královicová, S., and Brown, D.R. (2006). Hydrogen peroxide cleavage of the prion protein generates a fragment able to initiate polymerisation of full length prion protein. Int J Biochem Cell Biol. *38*,1429-1440.

Barnham, K.J., and Bush, A.I. (2008). Metals in Alzheimer's and Parkinson's diseases. Curr. Opin. Chem. Biol. 12, 222-228.

Basu, S., Mohan, M.L., Luo, X., Kundu, B., Kong, Q., and Singh, N. (2007). Modulation of PK-resistant PrP^{Sc} in cells and infectious brain homogenate by redox-iron: Implications for prion replication and disease pathogenesis. Mol. Biol. Cell. *18*, 3302-3312.

Beard, J.L., and Connor, J.R. (2003). Iron status and neural functioning. Ann. Rev. Nutr. *23*, 41-58.

Bocharova, O.V., Breydo, L., Salnikov, V.V., and Baskakov, I.V. (2005) Copper(II) inhibits *in vitro* conversion of prion protein into amyloid fibrils. Biochem. 44, 6776-6787.

Bounhar, Y., Zhang, Y., Goodyer, C.G,, and LeBlanc, A. (2001). Prion protein protects human neurons against Bax-mediated apoptosis. J. Biol. Chem. 276, 39145-39149.

Brazier, M.W., Davies, P., Player, E., Marken, F., Viles, J.H., and Brown, D.R. (2008). Manganese binding to the prion protein. J. Biol. Chem. *283*, 12831-12839.

Brown, D.R, Qin, K., Herms, J.W., Madlung, A., Manson, J., Strome, R., Fraser, P.E., Kruck, T., Von Bohlen, A., Schulz-Schaeffer, W., Giese, A., Westaway, D., and Kretzschmar, H. (1997). The cellular prion protein binds copper *in vivo*. Nature. *390*, 684-687.

Brown, D.R., Hafiz, F., Glasssmith, L.L., Wong, B.S., Jones, I.M., Clive, C., and Haswell, S.J. (2000). Consequences of manganese replacement of copper for prion protein function and proteinase resistance, EMBO J. *19*, 1180-1186.

Brown, D.R., Clive, C., and Haswell, S.J. (2001). Antioxidant activity related to copper binding of native prion protein. J. Neurochem. *76*, 69-76.

Brown, D.R., Schmidt, B., and Kretzschmar, H.A. (1998). Effects of copper on survival of prion protein knockout neurons and glia. J. Neurochem. *70*, 1686-1693.

Brown, L.R., and Harris, D.A. (2003). Copper and zinc cause delivery of the prion protein from the plasma membrane to a subset of early endosomes and the Golgi. J. Neurochem. *87*, 353-363.

Burns, C.S., Aronoff-Spencer, E., Legname, G., Prusiner, S.B., Antholine, W.E., Gerfen, G.J., Peisach, J., and Millhauser, G.L. (2003). Copper coordination in the full-length, recombinant prion protein. Biochem. *42*, 6794-803.

Büeler, H., Fischer, M., Lang, Y., Bluethmann, H., Lipp, H.P., DeArmond, S.J., Prusiner, S.B., Aguet, M., and Weissmann, C. (1992). Normal development and behaviour of mice lacking the neuronal cell-surface PrP protein. Nature. *356*, 577-582.

Caughey, B., and Baron, G.S. (2006). Prions and their partners in crime. Nature *443*, 803-810.

Chesebro, B., Trifilo, M., Race, R., Meade-White, K., Teng, C., LaCasse, R., Raymond, L., Favara, C., Baron, G., Priola, S., Caughey, B., Masliah, E., and Oldstone, M. (2005)Anchorless prion protein results in infectious amyloid disease without clinical scrapie. Science *308*, 1435-1439.

Deleault, N.R., Harris, B.T., Rees, J.R., and Supattapone, S. (2007). Formation of native prions from minimal components *in vitro*. Proc. Natl. Acad. Sci. USA *104*, 9741-9746.

Fernaeus, S., and Land, T. (2005). Increased iron-induced oxidative stress and toxicity in scrapie-infected neuroblastoma cells. Neurosci. Lett. *382*, 133-136.

Fernaeus, S., Halldin, J., Bedecs, K., and Land, T. (2005). Changed iron regulation in scrapie-infected neuroblastoma cells. Brain Res. Mol. Brain Res. *133*, 266-273.

Gaasch, J.A., Lockman, P.R., Geldenhuys, W.J., Allen, D.D., and Van der Schyf, C.J. (2007). Brain iron toxicity: differential responses of astrocytes, neurons, and endothelial cells. Neurochem Res. *32*, 1196-1208.

Gerlach, M., Ben-Shachar, D., Riederer, P., and Youdim, M.B. (1994). Altered brain metabolism of iron as a cause of neurodegenerative diseases J Neurochem. *63*, 793-807.

Giese, A., Buchholz, M., Herms, J., and Kretzschmar, H.A. (2005). Mouse brain synaptosomes accumulate copper-67 efficiently by two distinct processes independent of cellular prion protein. J. Mol. Neurosci. *27*, 347-54.

Giese, A., Levin, J., Bertsch, U., and Kretzschmar, H.A. (2004). Effect of metal ions on de novo aggregation of full-length prion protein. Biochem. Biophys. Res. Commun. *320*, 1240-1246.

Golts, N., Snyder, H., Frasier, M., Theisler, C., Choi, P., and Wolozin, B. (2002). Magnesium Inhibits Spontaneous and Iron-induced Aggregation of α-Synuclein. J. Biol. Chem. *277*, 16116-16123.

Gu, Y., Verghese, S., Mishra, R.S., Xu, X., Shi, Y., and Singh, N. (2003). Mutant prion protein-mediated aggregation of normal prion protein in the endoplasmic reticulum: implications for prion propagation and neurotoxicity. J. Neurochem. *84*, 10-22.

Harris, D.A., and True, H.L. (2006). New insights into prion structure and toxicity. Neuron. *50*, 353-357.

Hayashi,T., Shoji, M., and Abe, K. (2006). Molecular mechanisms of ischemic neuronal cell death--with relevance to Alzheimer's disease. Curr Alzheimer Res. *3*, 351-358.

Hegde, R.S., Mastrianni, J.A., Scott, M.R., DeFea, K.A., Tremblay, P., Torchia, M., DeArmond, S.J., Prusiner, S.B., and Lingappa, V.R. (1998). A transmembrane form of the prion protein in neurodegenerative disease. Science. *279*, 827-834.

Hegde, R.S., Tremblay, P., Groth, D., DeArmond, S.J., Prusiner, S.B., and Lingappa, V.R. (1999). Transmissible and genetic prion diseases share a common pathway of neurodegeneration. Nature. *402*, 822-826.

Heller, U., Winklhofer, K.F., Heske, J., Reintjes, A., and Tatzelt, J. (2003). Post-translational import of the prion protein into the endoplasmic reticulum interferes with cell viability: a critical role for the putative transmembrane domain. J. Biol. Chem. *278*, 36139-36147.

Herms, J., Tings, T., Gall, S., Madlung, A., Giese, A., Siebert, H., Schürmann, P., Windl, O., Brose, N., and Kretzschmar, H. (1999). Evidence of presynaptic location and function of the prion protein. J. Neurosci. *19*, 8866-8875.

Hesketh, S., Sassoon, J., Knight, R., and Brown, D.R. (2008). Elevated manganese levels in blood and CNS in human prion disease. Mol. Cell Neurosci. *37*, 590-598.

Hesketh, S., Sassoon, J., Knight, R., Hopkins, J., and Brown, D.R. (2007). Elevated manganese levels in blood and central nervous system occur before onset of clinical signs in scrapie and bovine spongiform encephalopathy. J. Anim. Sci. *85*, 1596-1609.

Hijazi, N., Shaked, Y., Rosenmann, H., Ben-Hur, T., and Gabizon, R. (2003). Copper binding to PrP^C may inhibit prion disease propagation. Brain Res. *12*, 192-200.

Hornshaw, M.P., McDermott, J.R., and Candy, J.M. (1995a). Copper binding to the N-terminal tandem repeat regions of mammalian and avian prion protein. Biochem. Biophys. Res. Commun. *207*, 621-629.

Hornshaw, M.P., McDermott, J.R., Candy, J.M., and Lakey, J.H. (1995b). Copper binding to the N-terminal tandem repeat region of mammalian and avian prion protein: structural studies using synthetic peptides, Biochem. Biophys. Res. Commun. *214*, 993-999.

Jackson, G.S., Murray, I., Hosszu, L.L., Gibbs, N., Waltho, J.P., Clarke, A.R., and Collinge, J. (2001). Location and properties of metal-binding sites on the human prion protein, Proc. Natl. Acad. Sci. U. S. A. *98*, 8531-8535.

Jeffrey, M., Goodsir, C.M., Race, R.E., and Chesebro, B. (2004). Scrapie-specific neuronal lesions are independent of neuronal PrP expression. Ann. Neurol. *55*, 781-792.

Jeong, S.Y., and David, S. (2003). Glycosylphosphatidyl-inositol-anchored ceruloplasmin is required for iron efflux from cells in the central nervous system. J. Biol. Chem. *278*, 27144-27148.

Jin, T., Gu, Y., Zanusso, G., Sy, M., Kumar, A., Cohen, M., Gambetti, P., and Singh, N. (2000). The chaperone protein BiP binds to a mutant prion protein and mediates its degradation by the proteasome. J. Biol. Chem. *275*, 38699-38704.

Jobling, M.F., Huang, X., Stewart, L.R., Barnham, K.J., Curtain, C., Volitakis, I., Perugini, M., White, A.R., Cherny, R.A., Masters, C.L., Barrow, C.J., Collins, S.J., Bush, A.I., and Cappai, R. (2001). Copper and zinc binding modulates the aggregation and neurotoxic properties of the prion peptide PrP106-126. Biochem. *40*, 8073-84.

Jobling, M.F., Stewart, L.R., White, A.R., McLean, C., Friedhuber, A., Maher, F., Beyreuther, K., Masters, C.L., Barrow, C.J., Collins, S.J., and Cappai, R. (1999). The hydrophobic core sequence modulates the neurotoxic and secondary structure properties of the prion peptide 106-126. J. Neurochem. *73*, 1557-1565.

Jones, C.E., Abdelraheim, S.R., Brown, D.R., and Viles, J.H. (2004). Preferential Cu^{2+} coordination by His96 and His111 induces beta-sheet formation in the unstructured amyloidogenic region of the prion protein. J. Biol. Chem. *279*, 32018-32027.

Jones, C.E., Klewpatinond, M., Abdelraheim, S.R., Brown, D.R., and Viles, J.H. (2005). Probing copper^{2+} binding to the prion protein using diamagnetic nickel^{2+} and 1H NMR: the unstructured N terminus facilitates the coordination of six copper^{2+} ions at physiological concentrations, J. Mol. Biol. *346*, 1393-1407.

Kaplan, J. (2002). Mechanisms of cellular iron acquisition: another iron in the fire. Cell. *111*, 603-606.

Kenward, A.G., Bartolotti, L.J., Burns, C.S. (2007). Copper and zinc promote interactions between membrane-anchored peptides of the metal binding domain of the prion protein. Biochem. *46*, 4261-4271.

Kiachopoulos, S., Heske, J., Tatzelt, J., and Winklhofer, K.F. (2004). Misfolding of the prion protein at the plasma membrane induces endocytosis, intracellular retention and degradation. Traffic. *5*, 426-436.

Kim, B.H., Jun, Y.C., Jin, J.K., Kim, J.I., Kim, N.H., Leibold, E.A., Connor, J.R., Choi, E.K., Carp, R.I., and Kim, Y.S. (2007). Alteration of iron regulatory proteins (IRP1 and IRP2) and ferritin in the brains of scrapie-infected mice. Neurosci. Lett. *422*, 158-163.

Kim, N.H., Choi, J.K., Jeong, B.H., Kim, J.I., Kwon, M.S., Carp, R.I., and Kim, Y.S. (2005). Effect of transition metals (Mn, Cu, Fe) and deoxycholic acid (DA) on the conversion of PrPC to PrPres. FASEB J. *19*, 783-785.

Kim, N.H, Park, S.J., Jin, J.K., Kwon, M.S., Choi, E.K., Karp, R.I., and Kim, Y.S. (2000). Increased ferric iron content and iron-induced oxidative stress in the brains of scrapie-infected mice. Brain Res. *884*, 98-103.

Kretzschmar, H.A., Tings, T., Madlung, A., Giese, A., Herms, J. (2000). Function of PrP(C) as a copper-binding protein at the synapse. Arch. Virol. Suppl. *16*, 239-49.

Kristiansen, M., Messenger, M.J., Klohn, P.C., Brandner, S., Wadsworth, J.D., Collinge, J., and Tabrizi, S.J. (2005). Disease-related prion protein forms aggresomes in neuronal cells leading to caspase-activation and apoptosis. J. Biol. Chem. *280*, 38851-38861.

Kuczius, T., Buschmann. A., Zhang, W., Karch, H., Becker, K., Peters, G., and Groschup, M.H. (2004). Cellular prion protein acquires resistance to proteolytic degradation following copper ion binding. Biol. Chem. *385*, 739-747.

Lee, K.S., Raymond, L.D., Schoen, B., Raymond, G.J., Kett, L., Moore, R.A., Johnson, L.M., Taubner, L., Speare, J.O., Onwubiko, H.A., Baron, G.S., Caughey, W.S., and Caughey, B. (2007). Hemin interactions and alterations of the subcellular localization of prion protein. J. Biol. Chem. *282*, 36525-36533.

Liu, M., Yu, S., Yang, J., Yin, X., and Zhao, D. (2007). RNA and CuCl2 induced conformational changes of the recombinant ovine prion protein. Mol. Cell Biochem. *294*, 197-203.

Ma, J., and Lindquist, S. (2002). Conversion of PrP to a self-perpetuating PrP^{Sc}-like conformation in the cytosol. Science. *298*, 1785-1788.

Ma, J., Wollmann, R., and Lindquist, S. (2002). Neurotoxicity and neurodegeneration when PrP accumulates in the cytosol. Science. *298*, 1781-1785.

Mallucci, G., Dickinson, A., Linehan, J., Klohn, P.C., Brandner, S., and Collinge, J. (2003). Depleting neuronal PrP in prion infection prevents disease and reverses spongiosis. Science. *302*, 871-874.

Mishra, R.S., Basu, S., Gu, Y., Luo, X., Zou, W.Q., Mishra, R., Li, R., Chen, S.G., Gambetti, P., Fujioka, H., and Singh, N. (2004). Protease-resistant human prion protein and ferritin are cotransported across Caco-2 epithelial cells: implications for species barrier in prion uptake from the intestine. J. Neurosci. *24*, 11280-11290.

Mishra, R.S., Gu, Y., Bose, S., Verghese, S., Kalepu, S., Singh, N. (2002). Cell surface accumulation of a truncated transmembrane prion protein in Gerstmann-Straussler-Scheinker disease P102L. J. Biol. Chem. *277*, 24554-24561.

Miura, T., Sasaki, S., Toyama, A., and Takeuchi, H. (2005). Copper reduction by the octapeptide repeat region of prion protein: pH dependence and implications in cellular copper uptake. Biochem. *44*, 8712-8720.

Moos, T., and Morgan, E.H. (2004). The metabolism of neuronal iron and its pathogenic role in neurological disease. Ann. N. Y. Acad. Sci. *1012*, 14-26.

Orem, N.R., Geoghegan, J.C., Deleault, N.R., Kascsak, R., and Supattapone, S. (2006). Copper (II) ions potently inhibit purified PrP^{res} amplification. J. Neurochem. *96*, 1409-1415.

Ostrerova-Golts, N., Petrucelli, L., Hardy, J., Lee, J.M., Farer, M., and Wolozin, B. (2000). The A53T alpha-synuclein mutation increases iron-dependent aggregation and toxicity. J. Neurosci. *20*, 6048-6054.

Paik, S.R., Shin, H.J., and Lee, J.H. (2000). Metal-catalyzed oxidation of alpha-synuclein in the presence of Copper (II) and hydrogen peroxide. Arch. Biochem. Biophys. *378*, 269-277.

Pamplona, R., Naudí, A., Gavín, R., Pastrana, M.A., Sajnani, G., Ilieva, E.V., Del Río, J.A., Portero-Otín, M., Ferrer, I., and Requena, J.R. (2008). Increased oxidation, glycoxidation, and lipoxidation of brain proteins in prion disease. Free Radic Biol Med. *45*, 1159-1166.

Pan, K.M., Stahl, N., and Prusiner, S.B. (1992). Purification and properties of the cellular prion protein from Syrian hamster brain, Protein Sci. *1*, 1343-1352.

Pauly, P.C., and Harris, D.A. (1998). Copper stimulates endocytosis of the prion protein. J. Biol. Chem. *273*, 33107-33110.

Perera, W.S.S., and Hooper, N.M. (2001). Ablation of the metal ion-induced endocytosis of the prion protein by disease-associated mutation of the octarepeat region. Curr Biol. *11*, 519-523.

Petersen, R.B., Siedlak, S.L., Lee, H.G., Kim, Y.S., Nunomura, A., Tagliavini, F., Ghetti, B., Cras, P., Moreira, P.I., Castellani, R.J., Guentchev, M., Budka, H., Ironside, J.W., Gambetti, P., Smith, M.A., and Perry, G. (2005). Redox metals and oxidative abnormalities in human prion diseases. Acta Neuropathol. (Berl). *110*, 232-238.

Prusiner, S.B. (1998). Prions. Proc. Natl. Acad. Sci. U. S. A. *95*, 3363-3383.

Qin, K., Yang, Y., Mastrangelo, P., and Westaway, D. (2002) Mapping Cu(II) binding sites in prion proteins by diethyl pyrocarbonate modification and matrix-assisted laser desorption ionization-time of flight (MALDI-TOF) mass spectrometric footprinting. J. Biol. Chem. *277*, 1981-1990.

Quaglio, E., Chiesa, R., and Harris, D.A. (2001). Copper converts the cellular prion protein into a protease-resistant species that is distinct from the scrapie isoform. J. Biol. Chem. *276*, 11432-11438.

Rossi, L., Lombardo, M.F., Ciriolo, M.R., and Rotilio, G. (2004). Mitochondrial dysfunction in neurodegenerative diseases associated with copper imbalance. Neurochem Res. *29*, 493-504.

Roucou, X., and LeBlanc, A.C. (2005). Cellular prion protein neuroprotective function: implications in prion diseases. J. Mol. Med. *83*, 3-11.

Roucou, X., Gains, M., and LeBlanc, A.C. (2004). Neuroprotective functions of prion protein. J. Neurosci. Res. *75*, 153-161.

Roucou, X., Giannopoulos, P.N., Zhang, Y., Jodoin, J., Goodyer, C.G., LeBlanc, A. (2005). Cellular prion protein inhibits proapoptotic Bax conformational change in human neurons and in breast carcinoma MCF-7 cells. Cell Death Differ. *12*, 783-795.

Sigurdsson, E.M., Brown, D.R., Alim, M.A., Scholtzova, H., Carp, R., Meeker, H.C., Prelli, F., Frangione, B., and Wisniewski, T. (2003). Copper chelation delays the onset of prion disease. J. Biol. Chem. *278*, 46199-46202.

Singh, A., Isaac, A.O., Luo, X., Mohan, M.L., Cohen, M.L., Chen, F., Kong, Q., Bartz, J., and Singh, N. (2009a). Abnormal brain iron homeostasis in human and animal prion disorders. Plos Pathog. *5*, e1000336.

Singh, A., Mohan, M.L., Isaac, A.O., Luo, X., and Singh, N. (2009b). Prion protein modulates cellular iron metabolism: Implications for prion disease pathogenesis. PlosOne. *4*, e4468.

Singh, A., Kong, Q., Luo, X., Petersen, R.B., Meyerson, H., and Singh N. (2009c). Prion protein knock-out mice are iron deficient: A functional role for PrP in iron metabolism. Plos One. *4*, e6115.

Smith, M.A., Harris, P.L., Sayre, L.M., and Perry, G. (1997). Iron accumulation in Alzheimer disease is a source of redox-generated free radicals. Proc. Natl. Acad. Sci. U. S. A. *94*, 9866-9868.

Smith, D.G., Cappai R., and Barnham K.J. (2007). The redox chemistry of the Alzheimer's disease amyloid beta peptide. Biochim Biophys Acta.*1768*,1976-1990.

Solforosi, L., Criado, J.R., McGavern, D.B., Wirz, S., Sánchez-Alavez, M., Sugama, S., DeGiorgio, L.A., Volpe, B.T., Wiseman, E., Abalos, G., Masliah, E., Gilden, D., Oldstone, M.B., Conti, B., and Williamson, R.A. (2004). Cross-linking cellular prion protein triggers neuronal apoptosis *in vivo*. Science. *303*, 1514-1516.

Stuermer, C.A., and Plattner, H. (2005). The 'lipid raft' microdomain proteins reggie-1 and reggie-2 (flotillins) are scaffolds for protein interaction and signalling. Biochem. Soc. Symp. *72*, 109-118.

Takahashi, M., Ko, L.W., Kulathingal, J., Jiang, P., Sevlever, D., and Yen, S.H. (2007). Oxidative stress-induced phosphorylation, degradation and aggregation of alpha-synuclein are linked to upregulated CK2 and cathepsin D. Eur. J. Neurosci. *26*, 863-874.

Thackray, A.M., Knight, R., Haswell, S.J., Bujdoso, R., and Brown, D.R. (2002). Metal imbalance and compromised antioxidant function are early changes in prion disease. Biochem. J. *362*, 253-258.

Thompson, K.J., Shoham, S., and Connor, J.R. (2001). Iron and neurodegenerative disorders. Brain Res. Bull. *55*, 155-164.

Treiber, C., Simons, A., and Multhaup, G. (2006). Effect of copper and manganese on the de novo generation of protease-resistant prion protein in yeast cells. Biochem. *45*, 6674-6680.

Treiber, C., Thompsett, A.R., Pipkorn, R., Brown, D.R., Multhaup, G. (2007). Real-time kinetics of discontinuous and highly conformational metal-ion binding sites of prion protein. J. Biol. Inorg. Chem. *12*, 711-720.

Tsenkova, R.N., Iordanova, I.K., Toyoda, K., and Brown, D.R. (2004). Prion protein fate governed by metal binding. Biochem. Biophys. Res. Commun. *325*, 1005-1012.

Vassallo, N., and Herms, J. (2003). Cellular prion protein function in copper homeostasis and redox signalling at the synapse. J. Neurochem. *86*, 538-544.

Waggoner, D.J., Drisaldi, B., Bartnikas, T.B., Casareno, R.L., Prohaska, J.R., Gitlin, J.D., and Harris, D.A. (2000). Brain copper content and cuproenzyme activity do not vary with prion protein expression level. J. Biol. Chem. *275*, 7455-7458.

Waheed, A., Grubb, J.H., Zhou, X.Y., Tomatsu, S., Fleming, R.E., Costaldi, M.E., Britton, R.S., Bacon. B.R., and Sly, W.S. (2002). Regulation of transferrin-mediated iron uptake by HFE, the protein defective in hereditary hemochromatosis. Proc. Natl. Acad. Sci. U. S. A. *99*, 3117-3122.

Walter, E.D, Stevens, D.J., Visconte, M.P., and Millhauser, G.L. (2007). The prion protein is a combined zinc and copper binding protein: Zn^{2+} alters the distribution of Cu^{2+} coordination modes. J. Am. Chem. Soc. *129*, 15440-15441.

Wang, X, Wang, F., Arterburn, L., Wollmann, R., and Ma, J. (2006). The interaction between cytoplasmic prion protein and the hydrophobic lipid core of membrane correlates with neurotoxicity. J. Biol. Chem. *281*, 13559-13565.

Watt, N.T, and Hooper, N.M. (2003). The prion protein and neuronal zinc homeostasis. Trends Biochem. Sci. *28*, 406-410.

Wong, B.S., Brown, D.R., Pan, T., Whiteman, M., Liu, T., Gambetti, P., Olesik, J., Rubenstein, R., and Sy, M.S. (2001a). Oxidative impairment in scrapie-infected mice is associated with brain metals. perturbations and altered antioxidant activities. J. Neurochem. *79*, 689-698.

Wong, B.S., Chen, S.G., Colucci, M., Xie, Z., Pan, T., Liu, T., Li, R., Gambetti, P., Sy, M.S., and Brown, D.R. (2001b). Aberrant metal binding by prion protein in human prion disease. J. Neurochem. *78*, 1400-1408.

Wong, B.S., Liu, T., Li, R., Pan, T., Petersen, R.B., Smith, M.A., Gambetti, P., Perry, G., Manson, J.C., Brown, D.R., and Sy, M-S. (2001c). Increased levels of oxidative stress markers detected in the brains of mice devoid of prion protein, J. Neurochem. *76*, 565-572.

Yadavalli, R., Guttmann, R.P., Seward, T., Centers, A.P., Williamson, R.A., and Telling, G.C. (2004). Calpain-dependent endoproteolytic cleavage of PrP^{Sc} modulates scrapie prion propagation. J. Biol. Chem. *279*, 21948-21956.

Chapter 5

Targeting of the Prion Protein to the Cytosol: Mechanisms and Consequences

Margit Miesbauer, Angelika S. Rambold, Konstanze F. Winklhofer and Jörg Tatzelt

Abstract
Prion diseases are characterized by the conformational transition of the cellular prion protein (PrP^C) into an aberrant protein conformer, designated scrapie-prion protein (PrP^{Sc}). A causal link between protein misfolding and neurodegeneration has been established for a variety of neurodegenerative disease, such as Alzheimer's disease, Parkinson's disease and polyglutamine diseases, but there is an ongoing debate about the nature of the neurotoxic species and how non-native conformers can damage neuronal populations. PrP is normally imported into the endoplasmic reticulum (ER) and targeted to the outer leaflet of the plasma membrane via a glycosylphosphatidylinositol (GPI) anchor. However, several conditions, such as ER stress or some pathogenic mutations in the PrP gene, can induce the mislocalization of PrP in the cytosol, where it has a neurotoxic potential as demonstrated in cell culture and transgenic mouse models. In this review we focus on intrinsic factors and cellular pathways implicated in the import of PrP into the ER and its mistargeting to the cytosol. The findings summarized here not only reveal a complex regulation of the biogenesis of PrP, but also provide interesting new insight into toxic activities of pathogenic protein conformers and quality control pathways of ER-targeted proteins.

Introduction
Prion diseases are a group of transmissible neurodegenerative disorders including Creutzfeldt-Jakob disease (CJD) and Gerstmann-Sträussler-Scheinker syndrome (GSS) in humans, scrapie in sheep and goat, bovine spongiform encephalopathy (BSE) in cattle and chronic wasting disease (CWD) in free-ranging deer. A hallmark of prion diseases is the conversion of the cellular prion protein PrP^C into its misfolded isoform, termed PrP^{Sc}. Similarly to other neurodegenerative diseases, there is an ongoing debate whether folding intermediates, oligomers, amorphous or fibrillar aggregates are the neurotoxic species (rev. in (Caughey and Lansbury, 2003; Haass and Selkoe, 2007; Winklhofer et al., 2008). A unique feature of prion diseases is their transmissibility, which is also linked to PrP^{Sc} constituting the main component of infectious prions (reviewed in Aguzzi and Polymenidou, 2004; Chesebro, 2003; Collinge, 2001; Prusiner et al., 1998; Weissmann et al., 1996). In the majority of prion diseases there is a correlation between the accumulation of misfolded PrP, formation of infectious prions and neurodegeneration, however, there are some interesting exceptions. Transgenic mouse models revealed that several PrP mutants can induce neuronal cell death in the absence of infectious prion propagation (Baumann et al., 2007; Chiesa et al., 1998; Flechsig et al., 2003; Hegde et al., 1998; Li et al., 2007; Ma et al., 2002; Muramoto et al., 1997; Shmerling et al., 1998). *Vice versa* it has been shown that propagation of infectious prions can occur in the absence of clinical signs (Brandner et al., 1996; Chesebro et al., 2005; Mallucci et al., 2003). Consequently, it appears plausible that the infectious particle and the neurotoxic agent might be two distinct species/conformers.

In this review we focus on a specific class of pathogenic PrP conformers that are characterized by a toxic potential in the cytosolic compartment. We summarize our current knowledge about PrP domains and cellular pathways regulating the translocation of PrP into the ER and pathophysiological consequences of PrP accumulation in the cytosol.

PrP-mediated neurodegeneration: a cytosolic connection
Several studies indicated that PrP can acquire a neurotoxic potential when its import into the ER is partially or completely compromised. Employing *in vitro* models Lingappa and coworkers demonstrated that during import into the ER, PrP can attain two different transmembrane topologies with either the N- or C-terminal domain facing the cytosol (Yost et al., 1990) (see below). A transgenic mouse model revealed that preventing the import of PrP into the ER leads to the formation of a neurotoxic PrP species in the cytosol. Mice expressing a PrP mutant lacking the N-terminal ER targeting signal (cytoPrP) acquired severe ataxia due to rapid cerebellar granule neuron degeneration (Ma et al., 2002). A recent study supported this activity of cytoPrP and demonstrated that neurotoxicity of cytoPrP is also seen in different neuronal populations (Wang et al., 2009). Cytotoxic effects of cytoPrP were also observed in several mammalian cell culture models (Ma et al., 2002; Rambold et al., 2006; Rane et al., 2004). Further support for a toxic potential of cytosolically localized PrP was obtained in a yeast model. During post-translational targeting of PrP to the ER, PrP was missorted to the cytosol and interfered with yeast growth (Heller et al., 2003).

So far, mutations within the N-terminal signal sequence of PrP, which could affect the efficiency of ER import, have not been identified in patients suffering from prion diseases. However, the pathogenic PrP-W145Stop mutant, linked to Gerstmann-Sträussler-Scheinker syndrome in humans,

The Prion Protein 2010 Jörg Tatzelt (Ed.) Savanna Press, UK
ISBN: 978-0-9543335-2-2. savannapress.com

has been reported to localize to the cytosolic and nuclear compartment (Zanusso et al., 1999). A different study including another pathogenic mutant, PrP-Q160Stop, corroborated these findings and provided the first evidence that information in the C-terminal domain of PrP is necessary and sufficient for ER import (Heske et al., 2004) (see below).

Quality control of secretory proteins

As outlined above, alterations in protein folding, processing and/or degradation are common features of prion diseases as well as other neurodegenerative disorders, implying a role of quality control systems, such as molecular chaperones and the ubiquitin-proteasome pathway in the removal of neurotoxic protein conformers (rev. in (Ciechanover and Brundin, 2003; Morimoto, 2008). In the cytosol the ubiquitin-proteasome system (UPS) is the main pathway for regulated protein turnover (rev. in (Ciechanover et al., 2000; Hershko and Ciechanover, 1998; Pickart, 2001). Moreover, the proteasome mediates post-translational degradation of non-native proteins generated in the ER as part of a quality control system, designated ER-associated degradation (ERAD). This pathway involves recognition of non-native polypeptides by ER-resident chaperones and retrograde transport to the cytosol where proteasomal degradation occurs (rev. in (Ellgaard and Helenius, 2003; Meusser et al., 2005; Nakatsukasa and Brodsky, 2008). In case ERAD substrates accumulate in the ER lumen, intracellular signaling pathways are induced, collectively called the unfolded protein response (UPR) (rev. in (Malhotra and Kaufman, 2007; Ron and Walter, 2007). Only little is known about quality control pathways of GPI-anchored proteins, such as PrP. Interestingly, a pre-emptive, co-translocational quality control pathway was described recently, which operates before translocation into the ER is completed (Kang et al., 2006; Oyadomari et al., 2006). Regulated translocation could act as an early quality control step to prevent an overload of the ER with non-native proteins, specifically during an imbalance in cellular homeostasis. Indeed, it has been shown that PrP can be co-translocationally re-directed to proteasomal degradation during conditions of acute ER stress (Kang et al., 2006; Orsi et al., 2006).

Import of PrP into the endoplasmic reticulum: the role of PrP domains

The N-terminal ER signal sequence

The primary signal to target PrP to the ER is a N-terminal signal sequence (aa 1-22), which is recognized and bound by the signal recognition particle (SRP). SRP interaction with both the emerging polypeptide chain and the ribosome results in pausing of elongation and targeting of the ribosome-nascent chain complex to the Sec61 translocon. After binding of SRP to its receptor at the ER membrane, protein synthesis proceeds and transfer of the polypeptide chain into the ER lumen occurs in a co-translational manner (Gilmore et al., 1982a; Gilmore et al., 1982b; Meyer and Dobberstein, 1980; Walter and Blobel, 1980; Walter and Blobel, 1982; Walter et al., 1984).

ER import can be mediated by a remarkably diverse set of signal sequences, and it has been demonstrated for various secretory pathway proteins that translocation efficiency is regulated in a signal peptide sequence-specific manner (rev. in (Hegde and Bernstein, 2006; Martoglio and Dobberstein, 1998; von Heijne, 1985). Indeed, the sequence of the ER signal peptide has a significant impact on the import efficiency and the formation of topological isoforms of PrP. The first hint that PrP can be synthesized in topologies others than the GPI-anchored form emerged from *in vitro* studies (Yost et al., 1990). Two different topologies, termed ^{Ntm}PrP (N-terminus facing the ER lumen) or ^{Ctm}PrP (C-terminus facing the ER lumen) were described with the internal hydrophobic (HD) domain (amino acids 112 - 135) serving as a putative transmembrane domain. Mutations within the internal HD or the N-terminal signal peptide were found to alter the relative amount of ^{Ctm}PrP and ^{Ntm}PrP (Hegde et al., 1998; Kim et al., 2001; Stewart et al., 2001). Interestingly, increased synthesis of ^{Ctm}PrP has been shown to coincide with progressive neurodegeneration both in Gerstmann-Sträussler-Scheinker syndrome (GSS) patients with an A117V mutation and in transgenic mice carrying a triple mutation within the hydrophobic domain (AAA113,115,118VVV, in short AV3) (Hegde et al., 1998; Stewart et al., 2005). For a more detailed description of ^{Ctm}PrP and other neurotoxic PrP mutants, please see the review by Harris and colleagues in the same issue.

The role of the ER signal sequence and the HD in the ER import efficiency of PrP was further emphasized when we expressed mammalian PrP in *Saccharomyces cerevisiae*. Due to the relatively low hydrophobicity index (Kyte-Doolittle) of the signal peptide, PrP was targeted to the ER via a post-translational pathway in yeast cells . As a consequence, ER import was impaired and the accumulation of PrP in the cytosol interfered with cell viability. By using a more hydrophobic signal peptide or by deleting the HD, import of PrP into the ER as well as and yeast growth could be restored (Heller et al., 2003).

The internal hydrophobic domain (HD)

The internal HD (aa 112-135) of PrP was first described as a putative transmembrane domain (Lopez et al., 1990). Later it was shown that mutations within the HD can alter the relative amount of ^{Ctm}PrP and ^{Ntm}PrP (Hegde et al., 1998; Kim et al., 2001; Stewart et al., 2001). Moreover, different models revealed that this stretch of hydrophobic amino acids has additional properties: 1. Short peptides comprising amino acid residues 106-126 form fibrils *in vitro* (Tagliavini et al., 1993) and induce cell death in cultured cells (Brown et al., 1996; Deli et al., 2000; Forloni et al., 1993; Haïk et al., 2000). 2. In polarized cells the HD confers basolateral sorting in a dominant manner (Uelhoff et al., 2005). 3. Transgenic mice expressing PrP mutants lacking the HD (PrPΔHD) spontaneously develop a non-transmissible neurodegenerative disease (Baumann et al., 2007; Li et al., 2007; Shmerling et al., 1998). 4. The HD is linked to the toxic activity of cytosolically localized PrP (Rambold et al., 2006) (see below). Moreover, we have recently shown that the HD promotes dimer formation of PrP, which is associated with its stress-protective activity (Rambold et al., 2008).

The C-terminal GPI signal peptide

Biochemical studies using purified PrP^{Sc} revealed that PrP is modified with a glycosylphosphatidylinositol (GPI) anchor (Stahl et al., 1987). GPI modification occurs in the ER via removal of a C-terminal signal peptide and the replacement with a pre-formed GPI anchor at an acceptor amino acid,

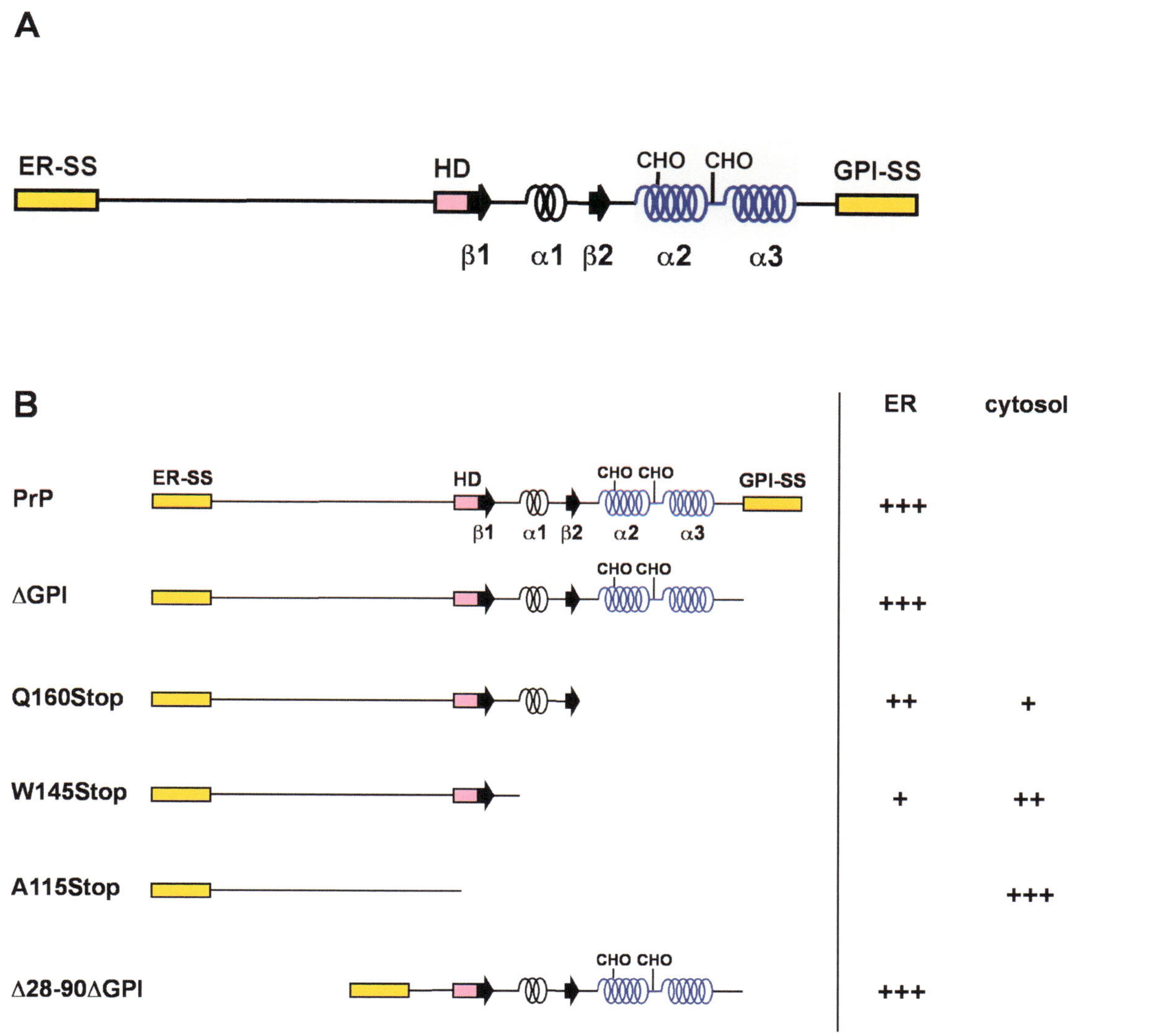

Figure 1. (A) Schematic presentation of the mammalian prion protein. ER-SS: ER signal sequence, α1-3: α-helical regions, β1, 2: β-strands, CHO: N-linked glycosylation acceptor site, GPI-SS: GPI anchor signal sequence, straight line: unstructured regions. Gray boxes mark domains implicated in the import of PrP into the ER. (B) The C-terminal structured domain is necessary and sufficient for effcient translocation into the ER. The scheme illustrates the ER import efficiency of different PrP mutants with deletions in the C- or N-terminal domain. While a deletion of the unstructured N-terminus and/or of the C-terminal GPI anchor signal sequence (PrPΔGPI, PrPΔNΔGPI) does not impair translocation of PrP the import efficiency successively decreased by increasing the deletions in the C-terminal structured domain.

called ω site. This reaction is catalyzed by a transamidase that resides within the ER membrane (Amthauer et al., 1993; Eisenhaber et al., 1998; Englund, 1993; Udenfriend and Kodukula, 1995). Similarly to the N-terminal signal peptides, the C-terminal GPI anchor signal sequences do not share conserved consensus sequences but rather characteristic features. Only a few amino acids with small side chains are allowed at the ω, the ω+1 and the ω+2 positions, and a hydrophobic domain is required about 10-12 amino acids distal to the ω site (Ferguson and Williams, 1988; Kodukula et al., 1993; Moran et al., 1991; Nuoffer et al., 1993; Nuoffer et al., 1991).

Deletion of the C-terminal GPI anchor signal sequence prevents membrane attachment of PrPΔGPI (also known as PrP-S230X or GPI$^-$PrP), but does not interfere with ER import or further trafficking of PrPΔGPI through the secretory pathway (Blochberger et al., 1997; Rogers et al., 1993; Winklhofer et al., 2003). PrPΔGPI is secreted both by cultured cells (Winklhofer et al., 2003) and by neurons of transgenic mice (Chesebro et al., 2005). However, membrane anchoring of PrP, be it via the authentic GPI anchor or via a heterologous CD4 transmembrane domain is required to ensure efficient modification of PrP with N-linked glycans of complex structure (Winklhofer et al., 2003).

Interestingly, the C-terminal GPI anchor signal peptide has been reported to mediate targeting of PrP to the ER. Initial results obtained *in vitro* indicated that the C-terminal signal peptide can promote post-translational translocation of PrP into the ER (Holscher et al., 2001). These findings were recently corroborated in cultured cells, moreover, it was shown that the GPI signal sequence can mediate co-translational ER targeting when located at the N-terminus of PrP or heterologous proteins (Gu et al., 2008).

The structured C-terminal domain

Structural studies on recombinant PrP indicated a particular modular composition (Figure 1A): the N-terminal domain of PrP spanning 120 amino acids is flexibly disordered followed by a highly structured C-terminal domain of approximately 110 amino acids. This autonomously folding domain contains three α-helical regions and a short, two-stranded β-sheet (Donne et al., 1997; Riek et al., 1996; Riek et al., 1997). The analysis of PrP-W145Stop, a pathogenic PrP mutant linked to Gerstmann-Sträussler-Scheinker syndrome in humans, provided the first hint that a loss of the C-terminal domain interferes with ER import of PrP: PrP-W145Stop was found in the cytosolic and nuclear compartment (Zanusso et al., 1999). Further studies including a different pathogenic mutant, PrP-Q160Stop, corroborated these findings and revealed that information in the C-terminal domain of PrP is necessary and sufficient for its import into the ER (Heske et al., 2004). We could now show that it is the presence of alpha-helical domains in the C-terminus that promotes ER import of PrP. The study revealed that after targeting to the translocon the secondary structure of the nascent polypeptide chain can significantly modulate translocation efficiency. ER-targeted polypeptides dominated by unstructured domains failed to efficiently translocate into the ER lumen and were subjected to proteasomal degradation via a co-translocational/pre-emptive pathway (Miesbauer et al., 2009).

Toxic PrP conformers in the cytosol

cytoPrP

A convincing model to demonstrate the neurotoxic capacity of cytosolically localized PrP (cytoPrP) was initially generated by Lindquist and coworkers: Mice expressing a PrP mutant with a deleted N-terminal ER targeting signal sequence acquire severe ataxia due to cerebellar degeneration (Ma et al., 2002). Importantly, in a different transgenic mouse model toxicity of cytoPrP was also demonstrated in other neuronal populations (Wang et al., 2009). Cell culture studies indicated that toxicity of cytoPrP seems to be dependent on its association with intracellular membranes (Wang et al., 2006) and its binding to Bcl-2, an anti-apoptotic protein present at the cytosolic side of ER and mitochondrial membranes (Rambold et al., 2006). Brain extracts from clinically ill transgenic mice expressing cytoPrP do not contain infectious prions, a phenomenon also seen in other transgenic mouse models of neurotoxic PrP mutants (rev. in (Winklhofer et al., 2008). In this context it should also be noted that cytosolically localized PrP seems to be present in a subset of neurons even under physiological conditions (Mironov et al., 2003).

PrP^{Sc}

The scrapie-prion protein or PrP^{Sc} was originally identified as a protease-resistant protein in brain extracts that co-purifies with the infectious scrapie agent (Bolton et al., 1982). In contrast to PrP^{C}, PrP^{Sc} is insoluble in non-ionic detergents, partially resistant to proteolytic digestion and characterized by a high content of β-sheet secondary structure (Caughey et al., 1991; Meyer et al., 1986; Pan et al., 1993). While PrP^{Sc} seems to be sufficient for the formation of infectious prions, elegant grafting experiments provided the first evidence that PrP^{Sc} is not toxic to neurons lacking PrP^{C} (Brandner et al., 1996), a finding later supported by conditional cell-type specific PrP knockout mice (Mallucci et al., 2003) and transgenic mice expressing PrPΔGPI, an anchorless PrP mutant (Chesebro et al., 2005). We have recently established a new cell culture model, which corroborates the observations from the mouse models and supports the notion that PrP^{Sc} is only toxic to cells expressing PrP^{C} (Rambold et al., 2008). Based on these findings, two plausible scenarios for the toxic effects of PrP^{Sc} can be envisaged, which are not mutually exclusive: Either neurotoxicity of PrP^{Sc} is linked to its propagation in neuronal cells, which is dependent on the expression of PrP^{C}, and/or PrP^{Sc} elicits a deadly signal through a PrP^{C}-dependent signaling pathway. Employing mouse models, cell culture and *in vitro* approaches, Tabrizi and colleagues provided support for the notion that during a prion infection misfolded PrP can translocate into the cytosol of the infected cells and induce toxic effects by inhibiting the proteasome (Kristiansen et al., 2007).

ER-targeted PrP can be re-directed to the cytosol via a co-translocational pathway

The cytosolic localization of ER proteins is usually linked to a cellular quality control system involving retrograde transport and subsequent proteasomal degradation, designated ER-associated degradation (ERAD). This pathway is designed to ensure cellular homeostasis and to preclude toxic effects of aberrant protein conformers within the ER. The accumulation of PrP in the cytosol after prolonged inhibition of proteasomal activity was initially interpreted as evidence for retrograde transport of PrP^{C} from the ER lumen into the cytosol (Ma and Lindquist, 2001; Yedidia et al., 2001). This conclusion, however, was discussed controversially after cytosolic accumulation of PrP has been proposed to be an experimental artifact, due to the prolonged treatment of cells with proteasomal inhibitors and overexpression of PrP driven by a viral promotor (Drisaldi et al., 2003). Based on our studies in yeast and mammalian cells we proposed that impaired ER import rather than retrograde translocation from the ER is responsible for the formation of cytosolic PrP (Heller et al., 2003; Heske et al., 2004; Tatzelt and Winklhofer, 2004) (Figure 1B). This concept was substantiated by recent studies showing that in addition to ERAD a co-translocational or pre-emptive quality control pathway exists, which interferes with the translocation of secretory pathway proteins into the ER lumen (Kang et al., 2006; Oyadomari et al., 2006). As a consequence, such a co-translocation quality control pathway reduces the amount of non-native substrates in the ER lumen and thereby unburdens the luminal ER quality control machinery. This could be particularly beneficial under conditions of ER stress in order to reduce the ER protein folding load (Kang et al., 2006).

Support for the idea that PrP is targeted via a co-translocational pathway to the cytosol was the observation that the cytosolic PrP fraction contained an uncleaved signal peptide (Heske et al., 2004; Miesbauer et al., 2009; Rambold et al., 2006; Zanusso et al., 1999). Processing of the N-terminal signal peptide and the transfer of core glycans occur co-translocationally prior to the release of the polypeptide into the ER lumen (rev. in (Kornfeld and Kornfeld, 1985). Thus, an uncleaved signal peptide is indicative for an abrogated translocation event. Interestingly,

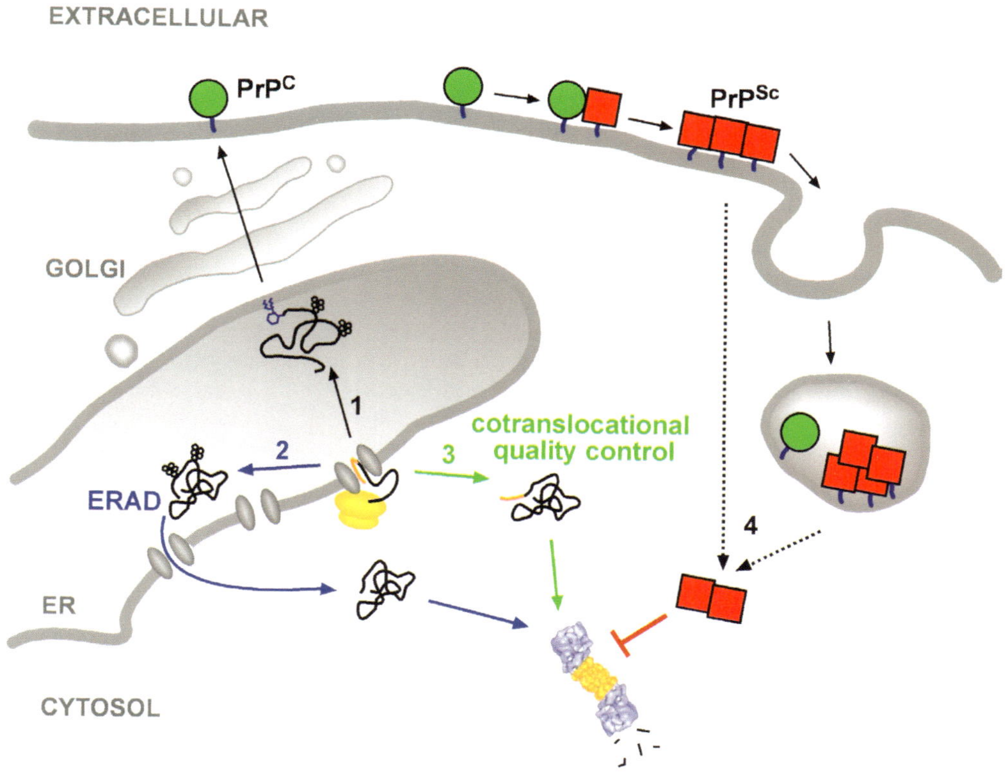

Figure 2. Pathways implicated in the fomation of cytosolically localized PrP. After targeting of the ribosome-nascent chain complex to the translocon the N-terminal signal peptide initiates the translocation process. 1: Under physiological conditions PrP is translocated into the ER lumen. During import the N-terminal signal peptide is cleaved off and two N-linked glycans are attached. Shortly after release into the ER lumen the C-terminal signal sequence is removed and replaced with a pre-formed GPI anchor. During transit through the secretory pathway the core glycans are converted into complex structures and PrP^{C} is finally targeted to the outer leaflet of the plasma membrane. In scrapie-infected cells, PrP^{C} at the cell surface or along the endocytic pathways interacts with and is converted into PrP^{Sc}. 2: The classical ERAD pathway would involve retro-translocation of non-native or misfolded conformers of PrP present in the ER lumen into the cytosol. The N-linked glycans are removed by a cytosolic N-glycanase prior to proteasomal degradation. 3: In a co-translocational quality control pathway productive translocation of the growing polypeptide chain into the ER is not pursued. Instead the protein is directly targeted to proteasomal degradation. Proteins subjected to this pathway would retain their N-terminal signal peptide. 4: In scrapie-infected cells oligomeric species of PrP^{Sc} might translocate from the plasma membrane or endocytic compartments to the cytosol where they interfere with proteasomal activity.

an uncleaved signal peptide was also described for ^{Ctm}PrP (Kim and Hegde, 2002; Stewart et al., 2001). Moreover, it has been shown that an uncleaved signal peptide is a characteristic feature of other substrates subjected to the co-translocational/pre-emptive pathway (Oyadomari et al., 2006; Rutkowski et al., 2007).

Why is it interesting that PrP seems to be subjected to a co-translocational quality control pathway? Little is known about quality control pathways of non-native or misfolded GPI-anchored proteins and ERAD in mammalian cells has mainly been studied with transmembrane proteins. Indeed, experimental evidence suggests that PrP is a poor ERAD substrate. In particular, PrP lacking a GPI anchor is misfolded and incompletely glycosylated, however, it is efficiently transported through the secretory pathway and finally secreted (Blochberger et al., 1997; Chesebro et al., 2005; Rogers et al., 1993; Walmsley et al., 2001; Winklhofer et al., 2003). This observation suggests that PrPΔGPI is not sensed as misfolded by the cellular quality control systems in the ER lumen. Two observations support such a hypothesis. Firstly, the terminal glucose residues are removed from the core glycans of PrPΔGPI and secondly, even the overexpression of PrPΔGPI does not result in the activation of the unfolded protein response (UPR) (Winklhofer et al., 2003). A recent study supports the idea that misfolded conformers of PrP generated in the ER lumen are neither recognized by ER quality control pathways nor routed to ERAD. Instead, mutant PrPs trafficked to the Golgi, from where the misfolded subpopulation was selectively trafficked for degradation in acidic compartments (Ashok and Hegde, 2009).

Which conditions might impair the import of PrP into the ER? Regarding the ER import mechanism, it is important to note that translocation of PrP into the ER seems to be more complex than that of other secretory proteins. *In vitro* studies using reconstituted lipid vesicles revealed that ER import of PrP requires the translocon-associated protein (TRAP) complex in addition to the essential translocon components Sec61 complex, the signal recognition particle receptor and the translocating chain associated protein (TRAM) (Fons et al., 2003). Consistent with the idea that an imbalance of cellular homeostasis brings a co-translocational quality control pathway on the scene (Rutkowski et al., 2007), the accumulation of cytosolic PrP containing an uncleaved signal peptide was favored under conditions of ER stress (Kang et al., 2006; Orsi et al., 2006).

Pathophysiological consequences of cytosolically localized PrP

A transgenic mouse model and various cell culture approaches provided convincing evidence that cytosolic PrP can have a toxic potential, however, two major questions have remained unanswered. First, does cytosolic PrP play a role in the pathogenesis of prion diseases and second, how does PrP in the cytosol induce neuronal cell death?

In this context it is important to remember that cytoPrP is biochemically quite different from infectious PrP^{Sc}. CytoPrP does adopt a misfolded and partially proteinase K (PK)-resistant conformation, however, it lacks post-translational modifications, like the GPI anchor and N-linked carbohydrates. Moreover, it does not seem to be infectious. A similar biochemical signature was described for PrPΔGPI. However, PrPΔGPI is efficiently imported into the ER and has no toxic activity in cultured cells (Rambold et al., 2006) nor does it cause neurodegeneration in transgenic mice (Chesebro et al., 2005). Based on these findings we established a cell culture model to identify cellular components linked to the toxic activity of cytoPrP (Rambold et al., 2006). Interestingly, an interaction of PrP and Bcl-2 was described previously for recombinantly expressed proteins (Kurschner and Morgan, 1995, 1996). We could verify such an interaction in neuronal cells and showed that the recruitment of Bcl-2 to misfolded PrP provides a plausible explanation for the toxic potential of cytoPrP (Rambold et al., 2006). A different study revealed that toxicity of cytoPrP seems to be dependent on its association with intracellular membranes (Wang et al., 2006), the locale where Bcl-2 resides. Regarding the role of Bcl-2 in cytoPrP-mediated toxicity, two scenarios are conceivable. Bcl-2 may be inactivated by its sequestration into cytoPrP aggregates. In support of this possibility, recent results from our group indicated that cytoPrP can displace Bax from a Bcl-2/Bax complex (Rambold *et al*, unpublished results). Alternatively, binding of Bcl-2 to cytoPrP may induce a conformational change of Bcl-2, converting it into a pro-apoptotic protein. Evidence for opposing phenotypes of Bcl-2 was provided by Lin *et al.*, who showed that the interaction with Nurr77 converts Bcl-2 from a protector to a killer protein (Lin et al., 2004).

Two additional findings need to be considered in the context of cytoPrP-induced toxicity. First, binding of Bcl-2 to PrP, as well as the toxic potential of cytoPrP could be mitigated by the increased expression of Hsp70 and its co-chaperone Hsp40, indicating a protective role of chaperones in PrP-induced toxicity (Rambold et al., 2006). Second, several cell culture models indicated that toxicity of cytoPrP is markedly enhanced under conditions of proteasomal impairment, suggesting that under physiological conditions the accumulation of cytoPrP is efficiently controlled (Ma et al., 2002; Rambold et al., 2006; Rane et al., 2004; Wang et al., 2005; Yedidia et al., 2001). This multifaceted influence on the formation and clearance of cytoPrP, which seems to be highly dependent on the cellular homeostasis, provides a plausible explanation for the discrepant findings in different reports on the toxic potential of cytosolic PrP (Fioriti et al., 2005; Roucou et al., 2003). In a different study it was shown that both ^{Ctm}PrP and cytoPrP can interact with and disrupt the function of Mahogunin, a cytosolic ubiquitin ligase whose loss causes spongiform neurodegeneration in mice (Chakrabarti and Hegde, 2009).

Interestingly, it was proposed that PrP^{Sc} might also exert its toxic potential in the cytosol.Tabrizi and colleagues showed that β-sheet-rich conformers of recombinant PrP and semi-purified PrP^{Sc} from prion-infected mouse brains inhibited the 26S proteasome *in vitro*. Moreover, proteasomal activity was decreased in prion-infected cells and in scrapie-diseased mouse brain (Kristiansen et al., 2007). This study not only might help to unravel pathophysiological mechanisms in prion diseases, it also supports the notion that oligomeric species rather than larger amyloid fibrils are the harmful species, a concept that was previously suggested for different neurodegenerative diseases (rev. in (Caughey and Lansbury, 2003; Haass and Selkoe, 2007). It will now be interesting to see how oligomeric PrP^{Sc} might translocate into the cytosol of scrapie-infected cells and how proteasomal impairment contributes to neuronal cell death in human prion diseases.

Future directions

To elucidate the mystery of a self-propagating protein conformer causing transmissible neurodegenerative diseases is a challenging mission, however, research on the prion protein has even more to offer. Transgenic mouse models paved the way for the intriguing concept that the infectious particle and the neurotoxic agent in prion diseases might be two distinct species. The existence of discrete conformers would also explain the phenomenon of sub-clinical prion infection (Hill et al., 2000). The question remains whether neurodegenerative diseases induced by neurotoxic PrP mutants in the absence of infectious prion propagation are genuine models for naturally occurring prion diseases. However, the established animal and cell culture models are extremely valuable to address an essential question in neurodegenerative disease research: How do aberrant protein conformers activate toxic signaling pathways in neuronal cells?

In addition to disease-related features, basic research on the prion protein provided new insight into cellular pathways of general interest. For example, studies employing PrP helped to reveal that the ribosome-nascent chain complex attached to the translocon complex is more dynamic than previously appreciated and that a co-translocational regulation of ER import provides an effcient mechanism to control loading of the ER with secretory pathway proteins. After all, understanding the cellular biology of disease-associated proteins is the essential prerequisite to exlore pathomechanistic events and to develop new therapeutic strategies.

Acknowledgements

Work of the authors was supported by grants from the Deutsche Forschungsgemeinschaft (SFB 596), the Max Planck Society, ForPrion, NeuroPrion and the BMBF.

References

Aguzzi, A., and Polymenidou, M. (2004). Mammalian prion biology: one century of evolving concepts. Cell *116*, 313-327.

Amthauer, R., Kodukula, K., Gerber, L., and Udenfriend, S. (1993). Evidence that the putative COOH-terminal signal transamidase involved in glycosylphosphatidylinositol protein synthesis is present in the endoplasmic reticulum. Proc Natl Acad Sci U S A *90*, 3973-3977.

Ashok, A., and Hegde, R.S. (2009). Selective processing and metabolism of disease-causing mutant prion proteins. PLoS Pathog *5*, e1000479.

Baumann, F., Tolnay, M., Brabeck, C., Pahnke, J., Kloz, U., Niemann, H.H., Heikenwalder, M., Rulicke, T., Burkle, A., and Aguzzi, A. (2007). Lethal recessive myelin toxicity of prion protein lacking its central domain. Embo J *26*, 538-547.

Blochberger, T.C., Cooper, C., Peretz, D., Tatzelt, J., Griffith, O.H., Baldwin, M.A., and Prusiner, S.B. (1997). Prion protein expression in Chinese hamster ovary cells using a glutamine synthetase selection and amplification system. Protein Eng *10*, 1465-1473.

Bolton, D.C., McKinley, M.P., and Prusiner, S.B. (1982). Identification of a protein that purifies with the scrapie prion. Science *218*, 1309-1311.

Brandner, S., Isenmann, S., Raeber, A., Fischer, M., Sailer, A., Kobayashi, Y., Marino, S., Weissmann, C., and Aguzzi, A. (1996). Normal host prion protein necessary for scrapie-induced neurotoxicity. Nature *379*, 339-343.

Brown, D.R., Schmidt, B., and Kretzschmar, H.A. (1996). Role of microglia and host prion protein in neurotoxicity of a prion protein fragment. Nature *380*, 345-347.

Caughey, B., and Lansbury, P.T. (2003). Protofibrils, pores, fibrils, and neurodegeneration: separating the responsible protein aggregates from the innocent bystanders. Annu Rev Neurosci *26*, 267-298.

Caughey, B.W., Dong, A., Bhat, K.S., Ernst, D., Hayes, S.F., and Caughey, W.S. (1991). Secondary structure analysis of the scrapie-associated protein PrP 27-30 in water by infrared spectroscopy. Biochemistry *30*, 7672-7680.

Chakrabarti, O., and Hegde, R.S. (2009). Functional depletion of mahogunin by cytosolically exposed prion protein contributes to neurodegeneration. Cell *137*, 1136-1147.

Chesebro, B. (2003). Introduction to the transmissible spongiform encephalopathies or prion diseases. British medical bulletin *66*, 1-20.

Chesebro, B., Trifilo, M., Race, R., Meade-White, K., Teng, C., LaCasse, R., Raymond, L., Favara, C., Baron, G., Priola, S., *et al.* (2005). Anchorless prion protein results in infectious amyloid disease without clinical scrapie. Science *308*, 1435-1439.

Chiesa, R., Piccardo, P., Ghetti, B., and Harris, D.A. (1998). Neurological illness in transgenic mice expressing a prion protein with an insertional mutation. Neuron *21*, 1339-1351.

Ciechanover, A., and Brundin, P. (2003). The ubiquitin proteasome system in neurodegenerative diseases: sometimes the chicken, sometimes the egg. Neuron *40*, 427-446.

Ciechanover, A., Orian, A., and Schwartz, A.L. (2000). Ubiquitin-mediated proteolysis: biological regulation via destruction. Bioessays *22*, 442-451.

Collinge, J. (2001). Prion diseases of humans and animals: their causes and molecular basis. Annu Rev Neurosci *24*, 519-550.

Deli, M.A., Sakaguchi, S., Nakaoke, R., Abraham, C.S., Takahata, H., Kopacek, J., Shigematsu, K., Katamine, S., and Niwa, M. (2000). PrP fragment 106-126 is toxic to cerebral endothelial cells expressing PrP(C). Neuroreport *11*, 3931-3936.

Donne, D.G., Viles, J.H., Groth, D., Mehlhorn, I., James, T.L., Cohen, F.E., Prusiner, S.B., Wright, P.E., and Dyson, H.J. (1997). Structure of the recombinant full-length hamster prion protein PrP(29-231): the N terminus is highly flexible. Proc Natl Acad Sci USA *94*, 13452-13457.

Drisaldi, B., Stewart, R.S., Adles, C., Stewart, L.R., Quaglio, E., Biasini, E., Fioriti, L., Chiesa, R., and Harris, D.A. (2003). Mutant PrP Is Delayed in Its Exit from the Neither Wild-type nor Mutant PrP Undergoes Proteasomal Degradation. J Biol Chem *278*, 21732-21743.

Eisenhaber, B., Bork, P., and Eisenhaber, F. (1998). Sequence properties of GPI-anchored proteins near the omega-site: constraints for the polypeptide binding site of the putative transamidase. Protein Eng *11*, 1155-1161.

Ellgaard, L., and Helenius, A. (2003). Quality control in the endoplasmic reticulum. Nat Rev Mol Cell Biol *4*, 181-191.

Englund, P.T. (1993). The structure and biosynthesis of glycosyl phosphatidylinositol protein anchors. Annu Rev Biochem *62*, 121-138.

Ferguson, M.A.J., and Williams, A.F. (1988). Cell-surface anchoring of protein via glycosyl-phosphatidylinositol structures. Annu Rev Biochem *57*, 285-320.

Fioriti, L., Dossena, S., Stewart, L.R., Stewart, R.S., Harris, D.A., Forloni, G., and Chiesa, R. (2005). Cytosolic prion protein (PrP) is not toxic in N2a cells and primary neurons expressing pathogenic PrP mutations. J Biol Chem *280*, 11320-11328.

Flechsig, E., Hegyi, I., Leimeroth, R., Zuniga, A., Rossi, D., Cozzio, A., Schwarz, P., Rulicke, T., Gotz, J., Aguzzi, A., *et al.* (2003). Expression of truncated PrP targeted to Purkinje cells of PrP knockout mice causes Purkinje cell death and ataxia. Embo J *22*, 3095-3101.

Fons, R.D., Bogert, B.A., and Hegde, R.S. (2003). Substrate-specific function of the translocon-associated protein complex during translocation across the ER membrane. J Cell Biol *160*, 529-539.

Forloni, G., Angeretti, N., Chiesa, R., Monzani, E., Salmona, M., Bugiani, O., and Tagliavini, F. (1993). Neurotoxicity of a prion protein fragment. Nature *362*, 543-546.

Gilmore, R., Blobel, G., and Walter, P. (1982a). Protein translocation across the endoplasmic reticulum. I. Detection in the microsomal membrane of a receptor for the signal recognition particle. J Cell Biol *95*, 463-469.

Gilmore, R., Walter, P., and Blobel, G. (1982b). Protein translocation across the endoplasmic reticulum. II. Isolation and characterization of the signal recognition particle receptor. J Cell Biol *95*, 470-477.

Gu, Y., Singh, A., Bose, S., and Singh, N. (2008). Pathogenic mutations in the glycosylphosphatidylinositol signal peptide of PrP modulate its topology in neuroblastoma cells. Mol Cell Neurosci *37*, 647-656.

Haass, C., and Selkoe, D.J. (2007). Soluble protein oligomers in neurodegeneration: lessons from the Alzheimer's amyloid beta-peptide. Nat Rev Mol Cell Biol *8*, 101-112.

Haïk, S., Peyrin, J.M., Lins, L., Rosseneu, M.Y., Brasseur, R., Langeveld, J.P., Tagliavini, F., Deslys, J.P., Lasmézas, C., and Dormont, D. (2000). Neurotoxicity of the Putative Transmembrane Domain of the Prion Protein. Neurobiol Dis *7*, 644-656.

Hegde, R.S., and Bernstein, H.D. (2006). The surprising complexity of signal sequences. Trends Biochem Sci *31*, 563-571.

Hegde, R.S., Mastrianni, J.A., Scott, M.R., DeFea, K.A., Tremblay, P., Torchia, M., DeArmond, S.J., Prusiner, S.B., and Lingappa, V.R. (1998). A transmembrane form of the prion protein in neurodegenerative disease. Science *279*, 827-834.

Heller, U., Winklhofer, K.F., Heske, J., Reintjes, A., and Tatzelt, J. (2003). Post-translational import of the prion protein into the endoplasmic reticulum interferes with cell viability: a critical role for the putative transmembrane domain. J Biol Chem *278*, 36139-36147.

Hershko, A., and Ciechanover, A. (1998). The ubiquitin system. Annu Rev Biochem *67*, 425-479.

Heske, J., Heller, U., Winklhofer, K.F., and Tatzelt, J. (2004). The C-terminal domain of the prion protein is necessary and sufficient for import into the endoplasmic reticulum. J Biol Chem *279*, 5435-5443.

Hill, A.F., Joiner, S., Linehan, J., Desbruslais, M., Lantos, P.L., and Collinge, J. (2000). Species-barrier-independent prion replication in apparently resistant species. Proc Natl Acad Sci U S A *97*, 10248-10253.

Holscher, C., Bach, U.C., and Dobberstein, B. (2001). Prion protein contains a second endoplasmic reticulum targeting signal sequence located at its C terminus. J Biol Chem *276*, 13388-13394.

Kang, S.W., Rane, N.S., Kim, S.J., Garrison, J.L., Taunton, J., and Hegde, R.S. (2006). Substrate-specific translocational attenuation during ER stress defines a pre-emptive quality control pathway. Cell *127*, 999-1013.

Kim, S.J., and Hegde, R.S. (2002). Cotranslational partitioning of nascent prion protein into multiple populations at the translocation channel. Mol Biol Cell *13*, 3775-3786.

Kim, S.J., Rahbar, R., and Hegde, R.S. (2001). Combinatorial control of prion protein biogenesis by the signal sequence and transmembrane domain. J Biol Chem *276*, 26132-26140.

Kodukula, K., Gerber, L.D., Amthauer, R., Brink, L., and Udenfriend, S. (1993). Biosynthesis of glycosylphosphatidylinositol (GPI)-anchored membrane proteins in intact cells: specific amino acid requirements adjacent to the site of cleavage and GPI attachment. J Cell Biol *120*, 657-664.

Kornfeld, R., and Kornfeld, S. (1985). Assembly of asparagine-linked oligosaccharides. Annu Rev Biochem *54*, 631-664.

Kristiansen, M., Deriziotis, P., Dimcheff, D.E., Jackson, G.S., Ovaa, H., Naumann, H., Clarke, A.R., van Leeuwen, F.W., Menendez-Benito, V., Dantuma, N.P., *et al.* (2007). Disease-associated prion protein oligomers inhibit the 26S proteasome. Mol Cell *26*, 175-188.

Kurschner, C., and Morgan, J.I. (1995). The cellular prion protein (PrP) selectively binds to Bcl-2 in the yeast two-hybrid system. Brain Res Mol Brain Res *30*, 165-168.

Kurschner, C., and Morgan, J.I. (1996). Analysis of interaction sites in homo- and heteromeric complexes containing Bcl-2 family members and the cellular prion protein. Brain Res Mol Brain Res *37*, 249-258.

Li, A., Christensen, H.M., Stewart, L.R., Roth, K.A., Chiesa, R., and Harris, D.A. (2007). Neonatal lethality in transgenic mice expressing prion protein with a deletion of residues 105-125. Embo J *26*, 548-558.

Lin, B., Kolluri, S.K., Lin, F., Liu, W., Han, Y.H., Cao, X., Dawson, M.I., Reed, J.C., and Zhang, X.K. (2004). Conversion of Bcl-2 from protector to killer by interaction with nuclear orphan receptor Nur77/TR3. Cell *116*, 527-540.

Lopez, C.D., Yost, C.S., Prusiner, S.B., Myers, R.M., and Lingappa, V.R. (1990). Unusual topogenic sequence directs prion protein biogenesis. Science *248*, 226-229.

Ma, J., and Lindquist, S. (2001). Wild-type PrP and a mutant associated with prion disease are subject to retrograde transport and proteasome degradation. Proc Natl Acad Sci U S A *98*, 14955-14960.

Ma, J., Wollmann, R., and Lindquist, S. (2002). Neurotoxicity and neurodegeneration when PrP accumulates in the cytosol. Science *298*, 1781-1785.

Malhotra, J.D., and Kaufman, R.J. (2007). The endoplasmic reticulum and the unfolded protein response. Semin Cell Dev Biol *18*, 716-731.

Mallucci, G., Dickinson, A., Linehan, J., Klohn, P.C., Brandner, S., and Collinge, J. (2003). Depleting neuronal PrP in prion infection prevents disease and reverses spongiosis. Science *302*, 871-874.

Martoglio, B., and Dobberstein, B. (1998). Signal sequences: more than just greasy peptides. Trends Cell Biol *8*, 410-415.

Meusser, B., Hirsch, C., Jarosch, E., and Sommer, T. (2005). ERAD: the long road to destruction. Nat Cell Biol *7*, 766-772.

Meyer, D.I., and Dobberstein, B. (1980). Identification and characterization of a membrane component essential for the translocation of nascent proteins across the membrane of the endoplasmic reticulum. J Cell Biol *87*, 503-508.

Meyer, R.K., McKinley, M.P., Bowman, K.A., Braunfeld, M.B., Barry, R.A., and Prusiner, S.B. (1986). Separation and properties of cellular and scrapie prion proteins. Proc Natl Acad Sci USA *83*, 2310-2314.

Miesbauer, M., Pfeiffer, N.V., Rambold, A.S., Muller, V., Kiachopoulos, S., Winklhofer, K.F., and Tatzelt, J. (2009). Alpha-helical domains promote translocation of intrinsically disordered polypeptides into the endoplasmic reticulum. J Biol Chem *284*, 24384-24393.

Mironov, A.J., Latawiec, D., Wille, H., Bouzamondo-Bernstein, E., Legname, G., Williamson, R.A., Burton, D., DeArmond, S.J., Prusiner, S.B., and Peters, P.J. (2003). Cytosolic prion protein in neurons. J Neurosci *23*, 7183-7193.

Moran, P., Raab, H., Kohr, W.J., and Caras, I.W. (1991). Glycophospholipid membrane anchor attachment. Molecular analysis of the cleavage/attachment site. J Biol Chem *266*, 1250-1257.

Morimoto, R.I. (2008). Proteotoxic stress and inducible chaperone networks in neurodegenerative disease and aging. Genes Dev *22*, 1427-1438.

Muramoto, T., DeArmond, S.J., Scott, M., Telling, G.C., Cohen, F.E., and Prusiner, S.B. (1997). Heritable disorder resembling neuronal storage disease in mice expressing prion protein with deletion of an a-helix. Nat Med *3*, 750-755.

Nakatsukasa, K., and Brodsky, J.L. (2008). The recognition and retrotranslocation of misfolded proteins from the endoplasmic reticulum. Traffic *9*, 861-870.

Nuoffer, C., Horvath, A., and Riezman, H. (1993). Analysis of the sequence requirements for glycosylphosphatidylinositol anchoring of Saccharomyces cerevisiae Gas1 protein. J Biol Chem *268*, 10558-10563.

Nuoffer, C., Jeno, P., Conzelmann, A., and Riezman, H. (1991). Determinants for glycophospholipid anchoring of the Saccharomyces cerevisiae GAS1 protein to the plasma membrane. Mol Cell Biol *11*, 27-37.

Orsi, A., Fioriti, L., Chiesa, R., and Sitia, R. (2006). Conditions of endoplasmic reticulum stress favor the accumulation of cytosolic prion protein. J Biol Chem *281*, 30431-30438.

Oyadomari, S., Yun, C., Fisher, E.A., Kreglinger, N., Kreibich, G., Oyadomari, M., Harding, H.P., Goodman, A.G., Harant, H., Garrison, J.L., *et al.* (2006). Cotranslocational degradation protects the stressed endoplasmic reticulum from protein overload. Cell *126*, 727-739.

Pan, K.-M., Baldwin, M., Nguyen, J., Gasset, M., Serban, A., Groth, D., Mehlhorn, I., Huang, Z., Fletterick, R.J., Cohen, F.E., *et al.* (1993). Conversion of a-helices into b-sheets features in the formation of the scrapie prion proteins. Proc Natl Acad Sci USA *90*, 10962-10966.

Pickart, C.M. (2001). Mechanisms underlying ubiquitination. Annu Rev Biochem *70*, 503-533.

Prusiner, S.B., Scott, M.R., DeArmond, S.J., and Cohen, F.E. (1998). Prion protein biology. Cell *93*, 337-348.

Rambold, A.S., Miesbauer, M., Rapaport, D., Bartke, T., Baier, M., Winklhofer, K.F., and Tatzelt, J. (2006). Association of Bcl-2 with misfolded prion protein is linked to the toxic potential of cytosolic PrP. Mol Biol Cell *17*, 3356-3368.

Rambold, A.S., Müller, V., Ron, U., Ben-Tal, N., Winklhofer, K.F., and Tatzelt, J. (2008). Stress-protective activity of prion protein is corrupted by scrapie-prions. Embo J *27*, 1974-1984.

Rane, N.S., Yonkovich, J.L., and Hegde, R.S. (2004). Protection from cytosolic prion protein toxicity by modulation of protein translocation. Embo J *23*, 4550-4559.

Riek, R., Hornemann, S., Wider, G., Billeter, M., Glockshuber, R., and Wuthrich, K. (1996). NMR structure of the mouse prion protein domain PrP(121-321). Nature *382*, 180-182.

Riek, R., Hornemann, S., Wider, G., Glockshuber, R., and Wuthrich, K. (1997). NMR characterization of the full-length recombinant murine prion protein, mPrP(23-231). FEBS Lett *413*, 282-288.

Rogers, M., Yehiely, F., Scott, M., and Prusiner, S.B. (1993). Conversion of truncated and elongated prion proteins into the scrapie isoform in cultured cells. Proc Natl Acad Sci U S A *90*, 3182-3186.

Ron, D., and Walter, P. (2007). Signal integration in the endoplasmic reticulum unfolded protein response. Nat Rev Mol Cell Biol *8*, 519-529.

Roucou, X., Guo, Q., Zhang, Y., Goodyer, C.G., and LeBlanc, A.C. (2003). Cytosolic prion protein is not toxic and protects against Bax-mediated cell death in human primary neurons. J Biol Chem *278*, 40877-40881.

Rutkowski, D.T., Kang, S.W., Goodman, A.G., Garrison, J.L., Taunton, J., Katze, M.G., Kaufman, R.J., and Hegde, R.S. (2007). The role of p58IPK in protecting the stressed endoplasmic reticulum. Mol Biol Cell *18*, 3681-3691.

Shmerling, D., Hegyi, I., Fischer, M., Blättler, T., Brandner, S., Götz, J., Rülicke, T., Flechsig, E., Cozzio, A., von Mehring, C., *et al.* (1998). Expression of animo-terminally truncated PrP in the mouse leading to ataxia and specific cerebellar lesions. Cell *93*, 203-214.

Stahl, N., Borchelt, D.R., Hsiao, K., and Prusiner, S.B. (1987). Scrapie prion protein contains a phosphatidylinositol glycolipid. Cell *51*, 229-240.

Stewart, R.S., Drisaldi, B., and Harris, D.A. (2001). A Transmembrane Form of the Prion Protein Contains an Uncleaved Signal Peptide and Is Retained in the Endoplasmic Reticululm. Mol Biol Cell *12*, 881-889.

Stewart, R.S., Piccardo, P., Ghetti, B., and Harris, D.A. (2005). Neurodegenerative illness in transgenic mice expressing a transmembrane form of the prion protein. J Neurosci *25*, 3469-3477.

Tagliavini, F., Prelli, F., Verga, L., Giaccone, G., Sarma, R., Gorevic, P., Ghetti, B., Passerini, F., Ghibaudi, E., Forloni, G., *et al.* (1993). Synthetic peptides homologous to prion protein residues 106-147 form amyloid-like fibrils in vitro. Proc Natl Acad Sci U S A *90*, 9678-9682.

Tatzelt, J., and Winklhofer, K.F. (2004). Folding and misfolding of the prion protein in the secretory pathway. Amyloid *11*, 162-172.

Udenfriend, S., and Kodukula, K. (1995). How glycosylphosphatidylinositol-anchored membrane proteins are made. Annu Rev Biochem *64*, 563-591.

Uelhoff, A., Tatzelt, J., Aguzzi, A., Winklhofer, K.F., and Haass, C. (2005). A pathogenic PrP mutation and doppel interfere with polarized sorting of the prion protein. J Biol Chem *280*, 5137-5140.

von Heijne, G. (1985). Signal sequences. The limits of variation. Journal of molecular biology *184*, 99-105.

Walmsley, A.R., Zeng, F.N., and Hooper, N.M. (2001). Membrane topology influences N-glycosylation of the prion protein. Embo J *20*, 703-712.

Walter, P., and Blobel, G. (1980). Purification of a membrane-associated protein complex required for protein translocation across the endoplasmic reticulum. Proc Natl Acad Sci U S A *77*, 7112-7116.

Walter, P., and Blobel, G. (1982). Signal recognition particle contains a 7S RNA essential for protein translocation across the endoplasmic reticulum. Nature *299*, 691-698.

Walter, P., Gilmore, R., and Blobel, G. (1984). Protein translocation across the endoplasmic reticulum. Cell *38*, 5-8.

Wang, X., Bowers, S.L., Wang, F., Pu, X.A., Nelson, R.J., and Ma, J. (2009). Cytoplasmic prion protein induces forebrain neurotoxicity. Biochim Biophys Acta.

Wang, X., Wang, F., Arterburn, L., Wollmann, R., and Ma, J. (2006). The interaction between cytoplasmic prion protein and the hydrophobic lipid core of membrane correlates with neurotoxicity. J Biol Chem *281*, 13559-13565.

Wang, X., Wang, F., Sy, M.S., and Ma, J. (2005). Calpain and other cytosolic proteases can contribute to the degradation of retro-translocated prion protein in the cytosol. J Biol Chem *280*, 317-325.

Weissmann, C., Fischer, M., Raeber, A., Büeler, H., Sailer, A., Shmerling, D., Rülicke, T., Brandner, S., and Aguzzi, A. (1996). The role of PrP in pathogenesis of experimental scrapie. Cold Spring Harb Symp Quant Biol *61*, 511-522.

Winklhofer, K.F., Heske, J., Heller, U., Reintjes, A., Muranji, W., Moarefi, I., and Tatzelt, J. (2003). Determinants of the in vivo-folding of the prion protein: a bipartite function of helix 1 in folding and aggregation. J Biol Chem *278*, 14961-14970.

Winklhofer, K.F., Tatzelt, J., and Haass, C. (2008). The two faces of protein misfolding: Gain and loss of function in neurodegenerative diseases. EMBO J *27*, 336-349.

Yedidia, Y., Horonchik, L., Tzaban, S., Yanai, A., and Taraboulos, A. (2001). Proteasomes and ubiquitin are involved in the turnover of the wild-type prion protein. Embo J *20*, 5383-5391.

Yost, C.S., Lopez, C.D., Prusiner, S.B., Myers, R.M., and Lingappa, V.R. (1990). Non-hydrophobic extracytoplasmic determinant of stop transfer in the prion protein. Nature *343*, 669-672.

Zanusso, G., Petersen, R.B., Jin, T., Jing, Y., Kanoush, R., Ferrari, S., Gambetti, P., and Singh, N. (1999). Proteasomal degradation and N-terminal protease resistance of the codon 145 mutant prion protein. J Biol Chem *274*, 23396-23404.

Chapter 6

The Role of GPI-anchored PrP^C in Mediating the Neurotoxic Effect of Scrapie Prions in Neurons

Helois E. Radford and Giovanna R. Mallucci

Abstract

There are two central phenomena in prion disease: prion replication and prion neurotoxicity. Underlying them both is the conversion of a host-encoded ubiquitously expressed protein, prion protein (PrP^C), into a partially-protease resistant isoform, PrP^{Sc}, which accumulates in the brain. PrP^{Sc} is associated with both pathology and infectivity (Prusiner, 1991). In the absence of PrP^C, PrP^{Sc} cannot be generated and PrP-null mice do not propagate infectivity or develop pathology on infection with scrapie (Bueler et al., 1993; Manson et al., 1994; Sailer et al., 1994). However, while PrP^C expression is fundamental to both prion infectivity and neurodegeneration, the uncoupling of these processes is a growing concept in the field. This dissociation is evident in subclinical states of prion infection (Hill et al., 2000), where PrP^{Sc} levels are high in the absence of disease, and in transgenic mice experiments, where, despite extra-neuronal PrP^{Sc} accumulation, even in very high amounts, there is no neurotoxicity (Chesebro et al., 2005; Mallucci et al., 2003). Both these models have further implications. Thus depleting PrP^C from neurons (but not glia) of prion-infected mice prevents clinical disease (Mallucci et al., 2003), and detaching it from the surface of cells by removing its anchor does the same (Chesebro et al., 2005). The uncoupling toxicity and infectivity has implications for the nature of the neurotoxic species; these mouse models suggest that the site for the generation of this species is intra-neuronal. This review considers the role of neuronal surface-expressed PrP^C in mediating toxicity in prion infection, and the dissociation of this from the deposition of PrP^{Sc}.

Introduction: concepts of neurotoxicity in prion disease

Prion diseases, or transmissible spongiform encephalopathies, are fatal neurodegenerative conditions of humans and animals, which are transmissible within or between mammalian species. They are associated with the accumulation in the brain of an abnormal, partially protease-resistant, isoform of host-encoded prion protein (PrP). The normal cellular isoform, PrP^C, is widely expressed but found at the highest level in the CNS as a glycosylated, glycosylphosphatidylinositol (GPI) anchored cell-surface protein (Caughey et al., 1989; Stahl et al., 1987).

The disease-related isoform (PrP^{Sc}) is derived from PrP^C by a post-translational process that involves conformational change and aggregation. PrP^C is rich in α-helical structure while PrP^{Sc}, which is found as insoluble aggregates, appears to be predominantly composed of ß-sheet structure. Many studies support the 'protein-only' hypothesis (Griffith, 1967; Prusiner, 1982) namely that an abnormal PrP isoform is the principal, and possibly the sole, constituent of the transmissible agent or prion. PrP^{Sc} is thought to act as a conformational template, recruiting PrP^C to form further PrP^{Sc}. Prion propagation may involve recruitment of an alternately folded form of PrP^C, into PrP^{Sc} aggregates, the process being driven thermodynamically by intermolecular interactions (Jackson et al., 1999).

However, the cause of neuronal death in prion disease remains unclear, in particular, the role played by PrP^{Sc}. The idea that neurodegeneration follows from direct toxicity of PrP^{Sc} and/or prions (Prusiner, 1991) has been increasingly challenged. PrP^{Sc} does not seem to have direct toxic effects on neurons *in vivo,* despite *in vitro* evidence for both full length PrP^{Sc} (Muller et al., 1993) and shorter fragments (Forloni et al., 1993) being toxic to primary cultured neurons. There are several inherited prion diseases in which PrP^{Sc} is not detected in significant amounts despite fatal clinical disease (Collinge et al., 1995; Medori et al., 1992; Tateishi et al., 1992), and often there is poor correlation between PrP^{Sc} deposits and clinical signs or neuronal loss (Collinge et al., 1995; Hsiao et al., 1990; Medori et al., 1992). Most tested potential therapeutic agents that target the accumulation of PrP^{Sc} have shown disappointing therapeutic benefit in prion infected animals. At best, they prolong incubation periods, even if given before, or at the time of, peripheral infection, but they do not prevent neurodegeneration and death (see (Mallucci and Collinge, 2005; Trevitt and Collinge, 2006) for reviews). What is clear is that host PrP^C is absolutely necessary for prion-induced neurotoxicity as well as for PrP^{Sc} production. This is seen in the resistance of PrP-null mice to prion neurotoxicity (Brandner et al., 1996; Bueler et al., 1993; Manson et al., 1994; Sailer et al., 1994), and in the rescue of neurons from degeneration when PrP^C is 'removed' from them, preventing ongoing neuronal prion replication (Mallucci et al., 2003; Mallucci et al., 2007; White et al., 2008) (see below). But even where PrP^C is present to support prion replication and mediate the neurotoxicity of PrP^{Sc}, PrP^{Sc} accumulation and toxicity can be uncoupled. This is seen in subclinical forms of prion infection in which animals can harbour high levels of infectivity, and PrP^{Sc}, without clinical disease (Hill and Collinge, 2003; Hill et al., 2000; Race et al., 2001), or where prion replication is extra-neuronal (Chesebro et al., 2005; Mallucci et al., 2003). In mice where neuronal PrP^C is deleted during the course of

The Prion Protein 2010 Jörg Tatzelt (Ed.) Savanna Press, UK
ISBN: 978-0-9543335-2-2. savannapress.com

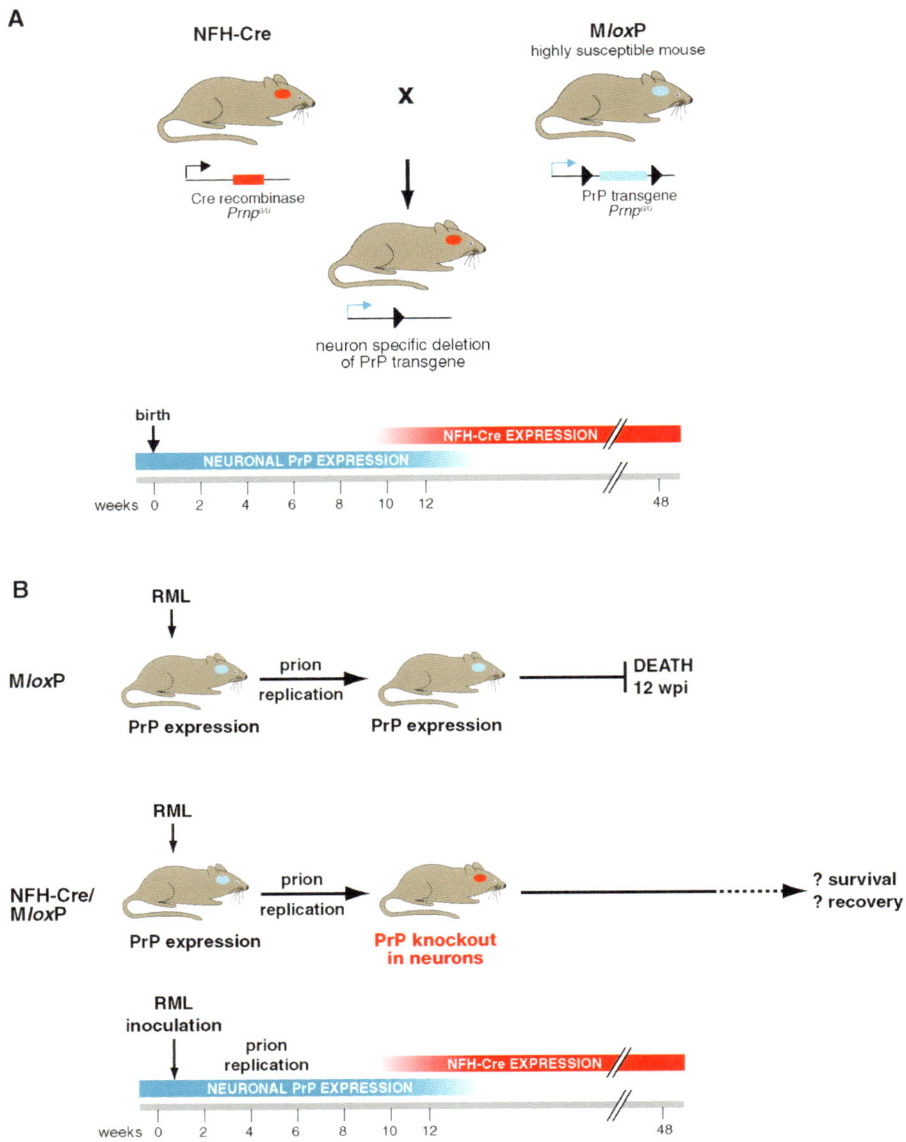

Figure 1. Adult onset PrP^{C} knockout in neurons by Cre-mediated recombination in mice; a model for studying neurotoxicity in prion disease **A.** NFH-Cre mice express Cre recombinase in neurons at ~ 9 weeks of age which deletes PrP transgenes from M*lox*P mice. **B.** The model is used to study the effects of neuronal PrP^{C} knockout during RML prion infection. Both NFH-Cre/M*lox*P and M*lox*P control mice replicate prions for 8 weeks, after which double transgenic animals lose neuronal PrP expression.

infection, but glial replication of PrP^{Sc} continues (Mallucci et al., 2003), and in mice in which the GPI-anchor is removed, detaching PrP from the cell surface but allowing replication to occur away from the cell bodies and processes (Chesebro et al., 2005), neurotoxicity does not occur despite extensive PrP^{Sc} accumulation.

One hypothesis is that if PrP^{Sc} itself is not directly neurotoxic, it is the depletion of PrP^{C} during prion replication that damages cells. The normal function of PrP^{C} is incompletely understood, and there is evidence that it may be neuroprotective (see review by Martins et al, this issue) and that aberrant trafficking or topology can be neurotoxic (Hegde et al., 1998; Hegde and Rane, 2003; Ma and Lindquist, 2002) (reviewed by Miesbauer et al; this issue). Yet the lack of neurodegeneration in either acquired (Mallucci et al., 2002), or embryonic (Bueler et al., 1993; Manson et al., 1994; Sailer et al., 1994), PrP knockout mice and in mice lacking correctly anchored (and hence functional) PrP (Chesebro et al., 2005), argues against this.

The uncoupling of PrP^{Sc} and neurotoxicity indicates that it is not the aggregated, protease resistant core of the molecule that is toxic to cells. But this begs the question of what the toxic species is? The central role of PrP^{C} conversion in the process suggests the generation of a toxic intermediate species, which may never aggregate, or whose toxicity is in any case independent of eventual aggregation. The possibility that this species may not only be transient and soluble, but needs to be generated within neurons is discussed below along with alternative explanations.

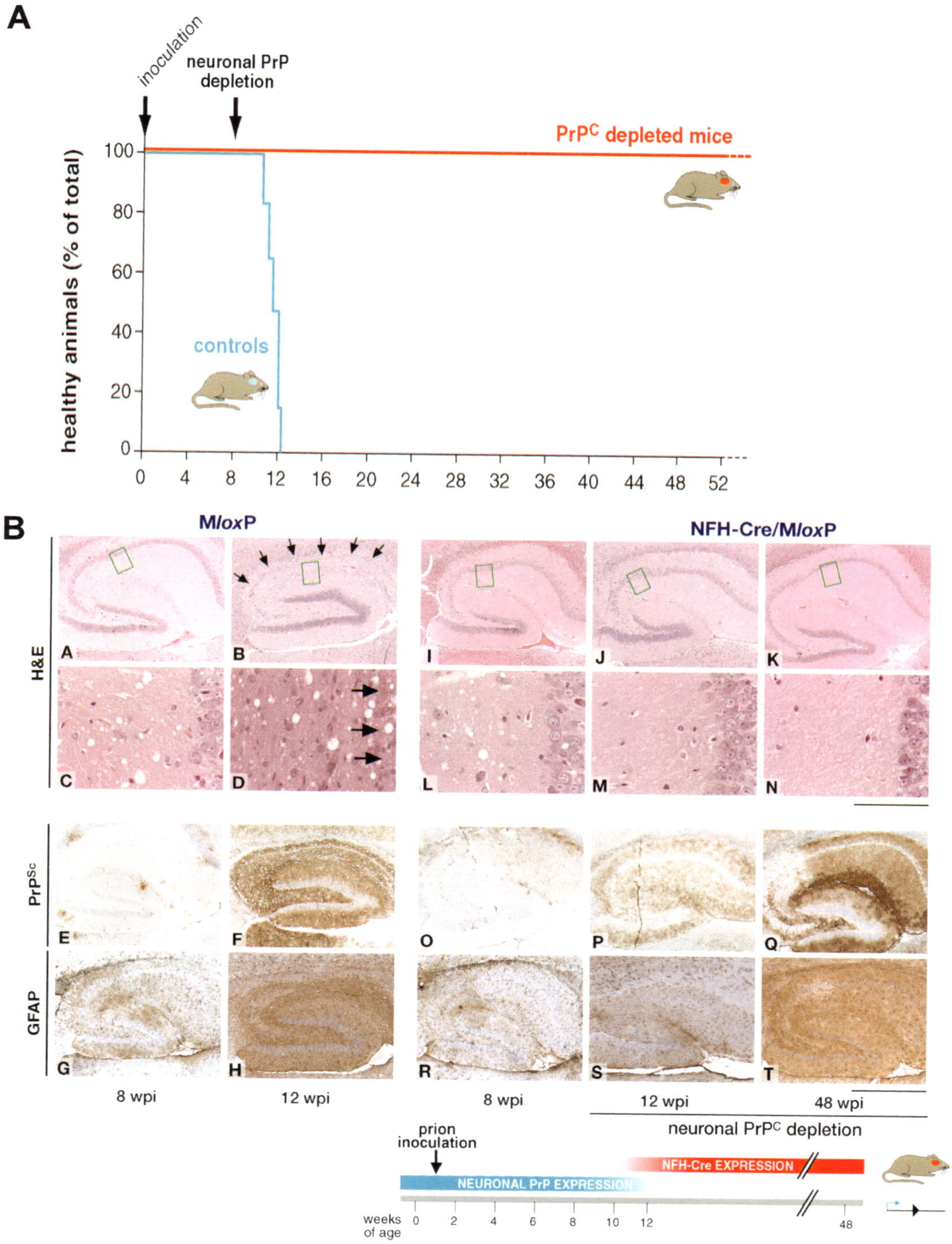

Figure 2. Scrapie-infected NFH-Cre/tg37 mice survive long term after Cre-mediated PrP^C depletion and early pathology reverses despite ongoing extra-neuronal PrP^{Sc} accumulation. **A.** Kaplan-Meier survival plot of RML scrapie prion-inoculated mice. All control M*lox*P mice (blue curve) succumbed to scrapie within 12 wpi (*n*=6). No animals with Cre-mediated PrP^C depletion at 8wpi (red line) succumbed to scrapie or show any clinical signs of disease by 52 wpi (*n*=6). The timing of inoculation and the onset of Cre-mediated PrP^C depletion 8 weeks into the course of infection, are indicated (arrows). **B.** Fixed sections of brains from scrapie-infected M*lox*P and NFH-Cre/M*lox*P mice, stained with haematoxylin and eosin (H&E), and immunostained for GFAP and PrP^{Sc} detection. Neuronal PrP depletion by Cre-mediated recombination in NFH-Cre/M*lox*P mice from 8 wpi is indicated. There is severe neuronal loss of hippocampal regions CA1-3 (arrows) (B,D) and indeed shrinkage of the entire hippocampus (B) in clinically sick M*lox*P mice, but no neuronal loss in prion-infected mice with Cre-mediated PrP^C depletion (I-K, L-N) up to 48 wpi. Early spongiform change is seen at 8 wpi in all mice (C,L), which recovers in scrapie-infected animals after Cre-mediated recombination (M,N). PrP^{Sc} accumulation (P,Q) and gliosis (S,T) continues in NFH-Cre/M*lox*P mice after Cre-mediated PrP^C depletion (R-T). Scale bar represents 320µm in all panels, except for panels C,D and L,M,N where it represents 80µm.

Prion neurotoxicity: a critical role for neuronal expression and surface anchoring of PrP^C?

The demonstration of recovery of early degenerative prion pathology and behavioral and neurophysiological deficits in mice in which neuronal PrP^C was depleted during infection, gives potentially important insights into possible mechanisms of prion neurotoxicity. The evidence comes from use of a mouse model in which PrP^C is depleted from neurons in the adult animal, after the nervous system is fully developed (Mallucci et al., 2002). Infecting these mice before knockout of PrP occurred allowed the effects of depleting neuronal PrP^C during the course of prion disease to be studied. In this system, PrP is expressed from 'floxed' transgenes (M*lox*P) and is deleted by expression of the enzyme Cre recombinase in neurons (from the NFH-Cre transgenes), see Figure 1. In contrast to PrP-null mice where deletion is embryonic, and global, the knockout of PrP here is acquired (at 9 weeks of age). Further, the knockout is neuron-specific, as the NFH promoter is expressed only in neurons. Mice were infected with prions at one week of age, 8 weeks before the onset of transgene mediated PrP depletion, allowing prion infection to develop over this time. By the time PrP depletion occurred, prion neuropathological change was established, with early spongiform change, astrocytosis and PrP^{Sc} deposition. Control M*lox*P animals without transgene meditated depletion at this stage of disease progressed to full prion neurodegeneration and death within 4 weeks. But animals in which PrP^C was knocked out at this time point survived, asymptomatic, long term (Figure 2A) and showed reversal of early spongiform change (Figure 2B, panels L,M,N). The animals were effectively clinically cured (Mallucci et al., 2003). This occurred despite the continued accumulation of extra-neuronal PrP^{Sc} over time to levels observed in terminally sick wild-type animals (Figure 2B, panel Q). This extra-neuronal conversion of PrP^C to PrP^{Sc} was not toxic to neurons, however. By 48 weeks post-infection, the animals had accumulated levels of PrP^{Sc} (and prion titres) as high as those seen in end-stage clinical prion disease in control mice, but no symptoms or neuronal loss were observed, suggesting that it is the occurrence of prion conversion within neurons that is pathogenic. Here, neurotoxicity and PrP^{Sc} accumulation are unequivocally uncoupled, and the data validated approaches in prion disease therapeutics targeting PrP^C; which was further confirmed using RNAi against PrP in prion infection (White et al., 2008). Further analysis of this model led to questions about the cell types involved in prion neurotoxicity, the nature of the toxic species and the timing of treatment in prion disease.

So does prion replication have to be intra-neuronal to be toxic? Certainly, this model supports this concept. Different cell types appear to be important in prion disease infectivity and pathogenesis. Both neurons and astrocytes propagate prions, and astrocytes may be the earliest sites of PrP^{Sc} accumulation (Diedrich et al., 1991). But astrocytic replication is not essential for prion neurotoxicity as expressing PrP in neurons alone in transgenic mice is sufficient to render the animals susceptible to prion disease (Race et al., 1995). Similarly, the data discussed above, where neuronal PrP^C depletion is protective against neurotoxicity, supports the argument that prion replication must be intra-neuronal to be toxic. Yet a mouse model in which PrP^C expression was directed towards astrocytes using the GFAP promoter to drive hamster PrP (HaPrP) expression also restored susceptibility to hamster scrapie prions in these mice (Raeber et al., 1997). These mice showed prion pathological changes (spongiform changes and astrogliosis), but no neuronal loss. Electron microscopy showed that neurons were primarily damaged and astrocytes appeared to not show any degenerative changes, although the $HaPrP^{Sc}$ mainly colocalised to astrocytes (Jeffrey et al., 2004). How does this reconcile with resistance of mice with acquired neuronal knockout of PrP to astrocyte PrP^{Sc} mediated neurotoxicity? The two models differ significantly and are not directly comparable for a number of reasons, including having been generated on different genetic backgrounds, expressing different species of PrP under different promoters and infection with different scrapie strains. GFAP-PrP mice over-express hamster PrP^C in astrocytes due to exponential induction of the GFAP promoter during infection, while in M*lox*P mice there is low-level astrocytic expression of mouse PrP^C expression under its own promoter. Further, 263K scrapie used to infect the GFAP-PrP mice produces very little neuronal loss, in contrast to the severe neurodegeneration induced by RML infection of M*lox*P mice. Further, it is increasingly clear that GFAP promoter activity is not entirely restricted to astrocytes. It is widely expressed in developing cerebellar interneurons, contributing up to 30% of the final population of these cells (Silbereis et al., 2009) and is also widely expressed throughout life in neural stem cells in the adult forebrain in the subventricular zone (Morshead et al., 2003). Recent evidence from genetic fate mapping of postnatal GFAP positive cells reveals that they seed the postnatal brain with neural progenitors that in turn give rise to their mature neuronal progeny throughout the CNS, including the cerebral cortex (Ganat et al., 2006). Thus it is likely that the GFAP-PrP mice are indeed expressing PrP in a sufficient number of neurons, to restore susceptibility of these mice to prion neurotoxicity.

While this mouse model makes a strong case for prion neurotoxicity hinging on the intra-neuronal conversion of PrP^C to PrP^{Sc}, it is also possible that in fact this need for neuronal replication actually reflects a kinetic effect rather than cell specificity. Thus the effect of gene dosage, and hence levels of expression, of PrP are known to correlate inversely with prion incubation times, despite extensive build-up of PrP^{Sc} and prion titre (Bueler et al., 1994; Weissmann et al., 1994). After neuronal PrP knockout, total PrP levels are reduced (Mallucci et al., 2003; Mallucci et al., 2002; Mallucci et al., 2007) and neurons no longer propagate prions; glial prion replication taking over. Beyond the reversal of pathology and increased survival in this model, what is striking is that by 48 wpi (weeks post infection) levels of PrP^{Sc} and titres of prion infectivity are as high as in the terminally ill wild type controls. The key difference, apart from the localization of the PrP^{Sc} to astrocytes, is the rate at which this accumulation occurs. The levels of infectivity and prion replication in control mice rise from basal to end stage in 4 weeks, while in animals with neuronal PrP knockout, the same increase takes 40 weeks. It is therefore also possible that at a lower rate of production, neurons can 'clear' the toxic species, wherever this is produced.

Collinge and colleagues have proposed a model to explain a 'clearance theory' for PrP toxic species, which they term PrP^L (for PrP 'lethal'). During the conversion of PrP^C to PrP^{Sc}, the putative neurotoxic intermediate molecule, PrP^L might be formed (Hill and Collinge, 2003; Collinge and

Clarke, 2007). Such an intermediate form of PrP may be different from the infectious form of PrP aggregates, and be rapidly cleared or sequestered into large PrP^{Sc} aggregates. According to this model, the rate of neurodegeneration would depend on the level of steady state 'PrP^L', which could explain the uncoupling of prion titre and neurotoxicity seen in experimental models. Preventing the conversion of PrP^C into PrP^{Sc} should also prevent the generation of 'PrP^L', whereas targeting PrP^{Sc}, the end product of the conversion process, might increase the level of 'PrP^L'. This model is consistent with the data from neuronally PrP-deleted mice, where the knockdown would impact on steady state level of a neurotoxic intermediate.

Further analysis of the Cre-mediated PrP knockout animals certainly supports the generation of a toxic species that is transient and separate to PrP^{Sc}. There are lessons to be learnt specifically from the course of reversal of the early pathology in these mice. Analysis showed that early pathology correlated with early functional deficits that recovered rapidly when PrP was deleted, consistent with transient toxicity. Moreover, the earliest 'toxic' effects were functional, not degenerative, implying a disturbance of neuronal function upstream of neurodegeneration. Thus early spongiform change was associated with cognitive and behavioral deficits and impaired neurophysiological function that recovered post knockout (Mallucci et al., 2007). Coincident with earliest spongiform change, 8 wpi, mice lost the capacity for novel object recognition, a test of non-spatial memory, and showed significantly reduced species- specific burrowing behavior, both of which reflect hippocampal dysfunction. Neurophysiological examination revealed concurrent reduced synaptic responses in CA1 hippocampal neurons. Remarkably, within a week, soon after neuronal PrP depletion both cognitive/behavioral and synaptic deficits recovered and this recovery was sustained up to 30 wpi, in parallel with recovery of spongiform change and sustained neuroprotection. Further, the deficits occur before extensive PrP^{Sc} deposits accumulate and recover rapidly after PrP^C depletion, again supporting the concept that they are caused by a transient neurotoxic species, distinct from aggregated PrP^{Sc} (Mallucci et al., 2007). The behavioral changes can similarly be prevented by lentivirally-mediated knockdown of PrP^C using RNA interference (White et al., 2008).

Thus this model uncouples prion neurotoxicity and PrP^{Sc} and prion titre, strongly supports the generation of a transient toxic intermediate, and overall supports the concept that this generation needs to be intra-neuronal to manifest its neurotoxicity. More recent evidence also implicates the necessity of intra-neuronal, or at least neuronal surface prion replication for neurotoxicity. PrP^C (and PrP^{Sc}) is attached to the cell surface by a glycosyphosphatidylinositol (GPI) moiety added to its C-terminus during processing in the Golgi, which anchors it to the outer leaflet of the plasma membrane (Caughey et al., 1989; Stahl et al., 1987). PrP^C is localised to cholesterol-rich lipid rafts within the plasma membrane, in part mediated by the affinity of the GPI anchor for saturated lipid species (Kaneko et al., 1997; Madore et al., 1999; Taraboulos et al., 1995; Vey et al., 1996), although some studies have shown PrP^C association with rafts in an independent manner, determined by PrP^C's N-terminus (Baron and Caughey, 2003; Campana et al., 2007; Sanghera and Pinheiro, 2002; Walmsley et al., 2003). The function of GPI anchors is poorly understood, and GPI anchored proteins may serve a variety of functions (receptors, adhesion, enzymes, complement regulation, signal transduction). The enrichment of GPI anchors, along with glycosphingolipids and cholesterol, in rafts locates them within these relatively rigid platforms on which, through the interaction of surface and cytoplasmic proteins, several signal transduction pathways are triggered and vesicular trafficking is organized (Fontaine et al., 2003; Mukasa et al., 1995; Oxley and Bacic, 1999). Recent evidence also suggests that GPI-linkage is required for many proteins in order to assume their fully functional conformation (Macrae et al., 2005). GPI-anchorage might also be important for apical targeting of proteins in polarised cells such as those of the intestinal epithelium, which conceivably could apply also to neurons.

The role of the GPI anchor in PrP^C function is not clear, but is it necessary for the toxicity and replication of prions? In cell-free experiments and *in vitro* anchorless PrP^C can be converted to PrP^{Sc} (Caughey and Raymond, 1991; Kocisko et al., 1994; Lawson et al., 2001; Rogers et al., 1993). However, others report that removing the anchor from PrP^C by treating with phosphatidylinositol-specific phospholipase C (PI-PLC) before exposure to scrapie prevents both infection and propagation in N2a cells (Enari et al., 2001). PI-PLC removes the anchor from PrP^C but not from PrP^{Sc}, but cathepsin-D does cleave the GPI moiety from PrP^{Sc}. Cathespin-D treated RML brain homogenate can infect both N2a cells and wild type mice (Lewis et al., 2006), suggesting that the GPI anchor on PrP^{Sc} is not needed for infection.

Chesebro and co-workers generated mice expressing anchorless PrP^C to address the role of the anchor in prion replication and neurotoxicity *in vivo*. These animals produced anchorless soluble monomeric PrP^C (GPI^- PrP). In neurons cultured from these mice, GPI^- PrP was not detectable on the plasma membrane, instead, around 90% was secreted into the medium and the rest detected in the ER and Golgi (Campana et al., 2007; Chesebro et al., 2005). Thus GPI^- PrP is secreted rather than retained intracellulary, indeed intracellular levels of GPI^- PrP were similar to those in control mice (Chesebro et al., 2005). These animals were essentially healthy and viable, but most significantly, were resistant to prion disease. Animals expressing GPI^- PrP did not develop clinical prion disease symptoms when infected with various mouse prion strains (RML, ME7, 22L) up to 400-600 days post infection (dpi), compared to control wild type mice that developed clinical symptoms within 140-160 days, depending on the strain used (Chesebro et al., 2005). Strikingly, however, the animals had widespread amyloid plaques of PrP^{Sc} throughout the brain, particularly near blood vessels. Plaques were detected from around 70 dpi and were progressively more widely distributed after 213 dpi. Some animals had up to 40% more PrP^{Sc} than clinically sick infected non-transgenic controls. GPI^- PrP expression thus supports prion replication and PrP^{Sc} deposition, but not the development of neurotoxicity or clinical disease, reminiscent of neuronal PrP knockout mice (Figure 3). Again there is clear uncoupling of PrP^{Sc} accumulation and neurotoxicity, and the lack of surface anchoring of PrP^C in this model further supports a role for prion neuronal conversion, mediated by GPI-anchored PrP^C, as central to neurotoxicity. That it is PrP^C, and not PrP^{Sc}, that must be anchored is confirmed by the fact that blood or brain

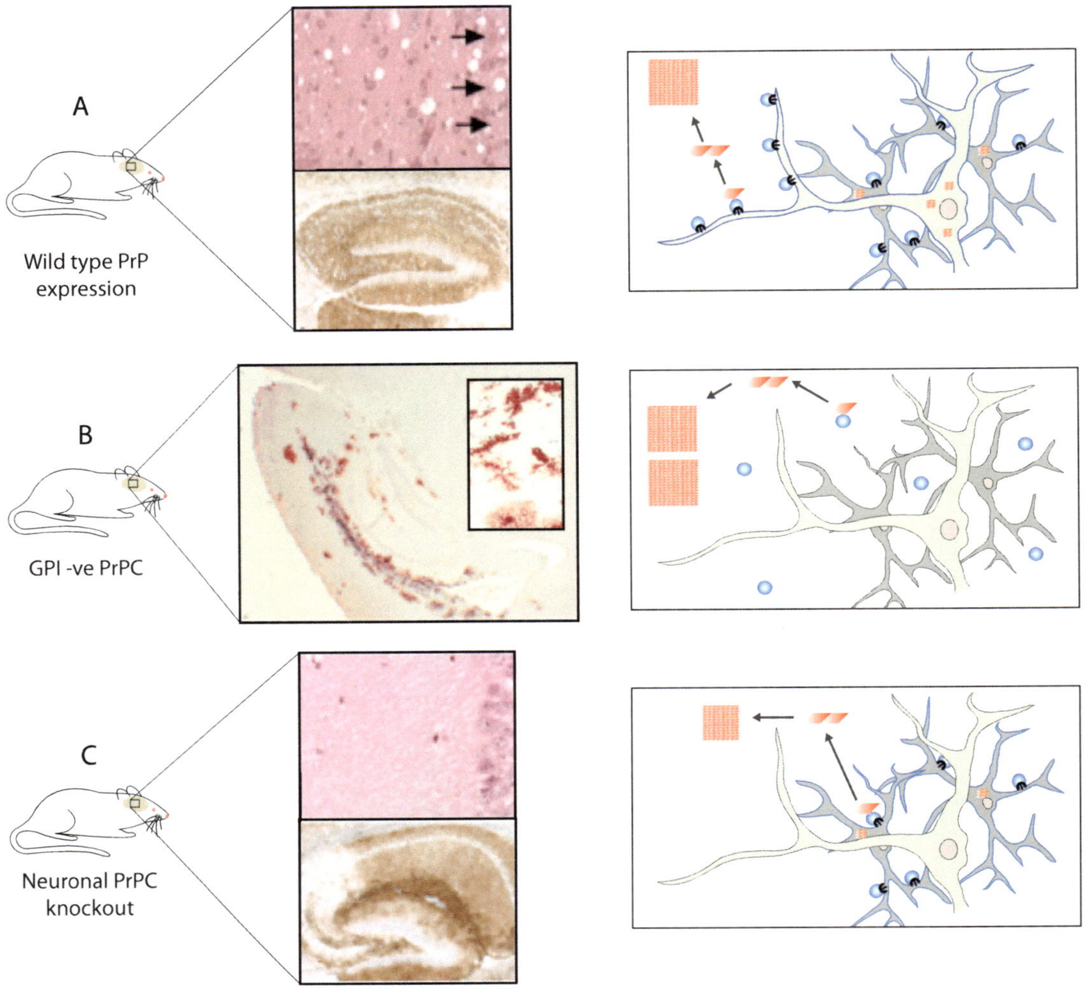

Figure 3. Uncoupling of PrP^{Sc} depostion and neurotoxicity in transgenic mice models. **A.** Mice expressing GPI-anchored PrP^{C} (blue circles = PrP^{C}; GPI anchor in black) physiologically on neurons (pale gray) and astrocytes (darker gray), replicate prions and deposit PrP^{Sc}, which aggregates (orange squares), causing typical prion neurotoxicity with neuronal loss and spongiosis (upper histology image) and PrP^{Sc} depostion (lower immunohistochemistry image). **B.** Mice expressing anchorless PrP^{C} (blue circles; no anchor) do not have PrP^{C} on the neuronal or astrocytic surface but secrete it, where it is converted into PrP^{Sc} and aggregates into amyloid plaques (histology panel). **C.** Mice with neuronal PrP knockout but expressing GPI-anchored PrP^{C} on astrocytes only generate PrP^{Sc} but are resistant to neurotoxicity (top histology panel). All mice generate PrP^{Sc} (lower immunohistochemistry panel) but the mice survive despite its accumulation, without neurotoxicity.

homogenates from RML infected GPI-negative mice were infectious to wild type mice (Chesebro et al., 2005; Trifilo et al., 2006), but not to other GPI^- PrP mice (Chesebro et al., 2005). The findings are consistent also with observations on the infectivity of cathepsin-D treated PrP^{Sc} (Lewis et al., 2006). Further, replacing the GPI anchor of PrP^{C} with its transmembrane domain, removes it from rafts and inhibits its conversion to PrP^{Sc} *in vitro* (Kaneko et al., 1997).

This work supports the hypothesis that the surface conversion of PrP^{C} to PrP^{Sc} requires PrP^{C}, but not PrP^{Sc}, to be GPI anchored. As neurons are the target of prion neurotoxicity it would appear that this process must occur on, or within, neurons. The data have been interpreted as reflecting prion neurotoxicity involving some perturbation of the normal function of PrP^{C}, possibly mediated through its anchor, in the presence of prion replication (Aguzzi, 2005). However, other interpretations are possible. The GPI^- PrP mouse model raises some questions. While they undoubtedly secrete soluble PrP, which is not located on the cell membrane and is involved in prion conversion extra-neuronally, these mice have rather low levels of mutant PrP expression. The authors describe levels of mRNA at about 50% wild type, but levels of actual protein appear lower still (see Chesebro et al; 2005, Figure S2). Gene dosage of PrP is a clear determinant of prion incubation time (Bueler et al., 1994; Manson et al., 1994; Weissmann et al., 1994) and even hemizygosity for wild type PrP leads to at least doubling of incubation time despite accumulating levels of PrP^{Sc} and prion titre. The reported survival of the GPI^- PrP mice is within this range, so is the survival unrelated to the anchoring but dependent on total levels of expression? It is likely that both the levels of PrP expression and the lack of anchor are significant. There is growing evidence in prion infected cell lines, in which protein trafficking has been selectively impaired, that the endosomal recycling compartment is the likely site of prion conversion (Marijanovic et al., 2009). This compartment is also known to be important in the pathway for internalization of GPI anchored proteins (Mayor et al., 1998), again supporting the need for a GPI anchor on PrP^{C} as being central to prion conversion, as anchorless PrP^{Sc} is infectious (Chesebro et al., 2005; Lewis et al., 2006).

Concluding remarks

The data discussed here unequivocally uncouple neurotoxicity and prion (and PrP^{Sc}) replication in prion disease, a concept described as early as 1994, when the effects of PrP gene dosage on incubation times were first described (Bueler et al., 1994; Weissmann et al., 1994). They also increasingly provide support for intra-neuronal generation of a toxic intermediate, where GPI-anchored PrP^{C} likely plays a crucial role. There are still many unanswered questions, not least what the neurotoxic species is and how, exactly, it causes neuronal dysfunction, and eventually death. The role of pharmacokinetic factors in clearing, depositing and aggregating PrP^{Sc} and its isoforms further complicates the interpretation of mechanisms of neurotoxicity in these disorders, but clearly is essential to an understanding of the real picture. Targeting the substrate of conversion, PrP^{C}, remains a therapeutically sound strategy, and new approaches may eventually provide clarity on the details of the toxic culprit and how it acts.

References

Aguzzi, A. (2005). Cell biology. Prion toxicity: all sail and no anchor. Science *308*, 1420- 1421.

Baron, G.S., and Caughey, B. (2003). Effect of glycosyl-phosphatidylinositol anchor-dependent and -independent prion protein association with model raft membranes on conversion to the protease-resistant isoform. J Biol Chem *278*, 14883-14892.

Brandner, S., Isenmann, S., Raeber, A., Fischer, M., Sailer, A., Kobayashi, Y., Marino, S., Weissmann, C., and Aguzzi, A. (1996). Normal host prion protein necessary for scrapie-induced neurotoxicity. Nature *379*, 339-343.

Bueler, H., Aguzzi, A., Sailer, A., Greiner, R.A., Autenried, P., Aguet, M., and Weissmann, C. (1993). Mice devoid of PrP are resistant to scrapie. Cell *73*, 1339-1347.

Bueler, H., Raeber, A., Sailer, A., Fischer, M., Aguzzi, A., and Weissmann, C. (1994). High prion and PrP^{Sc} levels but delayed onset of disease in scrapie-inoculated mice heterozygous for a disrupted PrP gene. Mol Med *1*, 19-30.

Campana, V., Caputo, A., Sarnataro, D., Paladino, S., Tivodar, S., and Zurzolo, C. (2007). Characterization of the properties and trafficking of an anchorless form of the prion protein. J Biol Chem *282*, 22747-22756.

Caughey, B., Race, R.E., Ernst, D., Buchmeier, M.J., and Chesebro, B. (1989). Prion protein biosynthesis in scrapie-infected and uninfected neuroblastoma cells. J Virol *63*, 175-181.

Caughey, B., and Raymond, G.J. (1991). The scrapie-associated form of PrP is made from a cell surface precursor that is both protease- and phospholipase-sensitive. J Biol Chem *266*, 18217-18223.

Chesebro, B., Trifilo, M., Race, R., Meade-White, K., Teng, C., LaCasse, R., Raymond, L., Favara, C., Baron, G., Priola, S., *et al.* (2005). Anchorless prion protein results in infectious amyloid disease without clinical scrapie. Science *308*, 1435-1439.

Collinge, J., Palmer, M.S., Sidle, K.C., Gowland, I., Medori, R., Ironside, J., and Lantos, P. (1995). Transmission of fatal familial insomnia to laboratory animals. Lancet *346*, 569-570.

Collinge, J ., and Clarke, A. (2007). A general model of prion strains and their pathogenicity. Science *318,* 930-936.

Diedrich, J.F., Bendheim, P.E., Kim, Y.S., Carp, R.I., and Haase, A.T. (1991). Scrapie-associated prion protein accumulates in astrocytes during scrapie infection. Proc Natl Acad Sci USA *88*, 375-379.

Enari, M., Flechsig, E., and Weissmann, C. (2001). scrapie prion protein accumulation by scrapie-infected neuroblastoma cells abrogated by exposure to a prion protein antibody. Proc Natl Acad Sci USA *98*, 9295-9299.

Fontaine, T., Magnin, T., Melhert, A., Lamont, D., Latge, J.P., and Ferguson, M.A. (2003). Structures of the glycosylphosphatidylinositol membrane anchors from Aspergillus fumigatus membrane proteins. Glycobiology *13*, 169-177.

Forloni, G., Angeretti, N., Chiesa, R., Monzani, E., Salmona, M., Bugiani, O., and Tagliavini, F. (1993). Neurotoxicity of a prion protein fragment. Nature *362*, 543-546.

Ganat, Y.M., Silbereis, J., Cave, C., Ngu, H., Anderson, G.M., Ohkubo, Y., Ment, L.R., and Vaccarino, F.M. (2006). Early postnatal astroglial cells produce multilineage precursors and neural stem cells *in vivo*. J Neurosci *26*, 8609-8621.

Griffith, J.S. (1967). Self-replication and scrapie. Nature *215*, 1043-1044.

Hegde, R.S., Mastrianni, J.A., Scott, M.R., DeFea, K.A., Tremblay, P., Torchia, M., DeArmond, S.J., Prusiner, S.B., and Lingappa, V.R. (1998). A transmembrane form of the prion protein in neurodegenerative disease. Science *279*, 827-834.

Hegde, R.S., and Rane, N.S. (2003). Prion protein trafficking and the development of neurodegeneration. Trends Neurosci *26*, 337-339.

Hill, A.F., and Collinge, J. (2003). Subclinical prion infection. Trends Microbiol *11*, 578- 584.

Hill, A.F., Joiner, S., Linehan, J., Desbruslais, M., Lantos, P.L., and Collinge, J. (2000). Species-barrier-independent prion replication in apparently resistant species. Proc Natl Acad Sci USA *97*, 10248-10253.

Hsiao, K.K., Scott, M., Foster, D., Groth, D.F., DeArmond, S.J., and Prusiner, S.B. (1990). Spontaneous neuro-degeneration in transgenic mice with mutant prion protein. Science *250*, 1587-1590.

Jackson, G.S., Hosszu, L.L., Power, A., Hill, A.F., Kenney, J., Saibil, H., Craven, C.J., Waltho, J.P., Clarke, A.R., and Collinge, J. (1999). Reversible conversion of monomeric human prion protein between native and fibrilogenic conformations. Science *283*, 1935- 1937.

Jeffrey, M., Goodsir, C.M., Race, R.E., and Chesebro, B. (2004). Scrapie-specific neuronal lesions are independent of neuronal PrP expression. Ann Neurol *55*, 781-792.

Kaneko, K., Vey, M., Scott, M., Pilkuhn, S., Cohen, F.E., and Prusiner, S.B. (1997). COOH-terminal sequence of the cellular prion protein directs subcellular trafficking and controls conversion into the scrapie isoform. Proc Natl Acad Sci USA *94*, 2333-2338.

Kocisko, D.A., Come, J.H., Priola, S.A., Chesebro, B., Raymond, G.J., Lansbury, P.T., and Caughey, B. (1994). Cell-free formation of protease-resistant prion protein. Nature *370*, 471-474.

Lawson, V.A., Priola, S.A., Wehrly, K., and Chesebro, B. (2001). N-terminal truncation of prion protein affects both formation and conformation of abnormal protease-resistant prion protein generated *in vitro*. J Biol Chem *276*, 35265-35271.

Lewis, P.A., Properzi, F., Prodromidou, K., Clarke, A.R., Collinge, J., and Jackson, G.S. (2006). Removal of the glycosylphosphatidylinositol anchor from PrP(sc) by cathepsin D does not reduce prion infectivity. Biochem J *395*, 443-448.

Ma, J., and Lindquist, S. (2002). Conversion of PrP to a self-perpetuating PrP^{Sc}-like conformation in the cytosol. Science *298*, 1785-1788.

Macrae, J.I., Acosta-Serrano, A., Morrice, N.A., Mehlert, A., and Ferguson, M.A. (2005). Structural characterization of NETNES, a novel glycoconjugate in Trypanosoma cruzi epimastigotes. J Biol Chem *280*, 12201-12211.

Madore, N., Smith, K.L., Graham, C.H., Jen, A., Brady, K., Hall, S., and Morris, R. (1999). Functionally different GPI proteins are organized in different domains on the neuronal surface. EMBO J *18*, 6917-6926.

Mallucci, G., and Collinge, J. (2005). Rational targeting for prion therapeutics. Nat Rev Neurosci *6*, 23-34.

Mallucci, G., Dickinson, A., Linehan, J., Klohn, P.C., Brandner, S., and Collinge, J. (2003). Depleting neuronal PrP in prion infection prevents disease and reverses spongiosis. Science *302*, 871-874.

Mallucci, G.R., Ratte, S., Asante, E.A., Linehan, J., Gowland, I., Jefferys, J.G., and Collinge, J. (2002). Post-natal knockout of prion protein alters hippocampal CA1 properties, but does not result in neurodegeneration. EMBO J *21*, 202-210.

Mallucci, G.R., White, M.D., Farmer, M., Dickinson, A., Khatun, H., Powell, A.D., Brandner, S., Jefferys, J.G., and Collinge, J. (2007). Targeting cellular prion protein reverses early cognitive deficits and neurophysiological dysfunction in prion-infected mice. Neuron *53*, 325-335.

Manson, J.C., Clarke, A.R., McBride, P.A., McConnell, I., and Hope, J. (1994). PrP gene dosage determines the timing but not the final intensity or distribution of lesions in scrapie pathology. Neurodegeneration *3*, 331-340.

Marijanovic, Z., Caputo, A., Campana, V., and Zurzolo, C. (2009). Identification of an intracellular site of prion conversion. PLoS Pathog *5*, e1000426.

Mayor, S., Sabharanjak, S., and Maxfield, F.R. (1998). Cholesterol-dependent retention of GPI-anchored proteins in endosomes. EMBO J *17*, 4626-4638.

Medori, R., Montagna, P., Tritschler, H.J., LeBlanc, A., Cortelli, P., Tinuper, P., Lugaresi, E., and Gambetti, P. (1992). Fatal familial insomnia: a second kindred with mutation of prion protein gene at codon 178. Neurology *42*, 669-670.

Morshead, C.M., Garcia, A.D., Sofroniew, M.V., and van Der Kooy, D. (2003). The ablation of glial fibrillary acidic protein-positive cells from the adult central nervous system results in the loss of forebrain neural stem cells but not retinal stem cells. Eur J Neurosci *18*, 76-84.

Mukasa, R., Umeda, M., Endo, T., Kobata, A., and Inoue, K. (1995). Characterization of glycosylphosphatidylinositol (GPI)-anchored NCAM on mouse skeletal muscle cell line C2C12: the structure of the GPI glycan and release during myogenesis. Arch Biochem Biophys *318*, 182-190.

Muller, W.E., Ushijima, H., Schroder, H.C., Forrest, J.M., Schatton, W.F., Rytik, P.G., and Heffner-Lauc, M. (1993). Cytoprotective effect of NMDA receptor antagonists on prion protein (PrionSc)-induced toxicity in rat cortical cell cultures. Eur J Pharmacol *246*, 261-267.

Oxley, D., and Bacic, A. (1999). Structure of the glycosylphosphatidylinositol anchor of an arabinogalactan protein from Pyrus communis suspension-cultured cells. Proc Natl Acad Sci USA *96*, 14246-14251.

Prusiner, S.B. (1982). Novel proteinaceous infectious particles cause scrapie. Science *216*, 136-144.

Prusiner, S.B. (1991). Molecular biology of prion diseases. Science *252*, 1515-1522.

Race, R., Raines, A., Raymond, G.J., Caughey, B., and Chesebro, B. (2001). Long-term subclinical carrier state precedes scrapie replication and adaptation in a resistant species: analogies to bovine spongiform encephalopathy and variant Creutzfeldt-Jakob disease in humans. J Virol *75*, 10106-10112.

Race, R.E., Priola, S.A., Bessen, R.A., Ernst, D., Dockter, J., Rall, G.F., Mucke, L., Chesebro, B., and Oldstone, M.B. (1995). Neuron-specific expression of a hamster prion protein minigene in transgenic mice induces susceptibility to hamster scrapie agent. Neuron *15*, 1183-1191.

Raeber, A.J., Race, R.E., Brandner, S., Priola, S.A., Sailer, A., Bessen, R.A., Mucke, L., Manson, J., Aguzzi, A., Oldstone, M.B., *et al.* (1997). Astrocyte-specific expression of hamster prion protein (PrP) renders PrP knockout mice susceptible to hamster scrapie. EMBO J *16*, 6057-6065.

Rogers, M., Yehiely, F., Scott, M., and Prusiner, S.B. (1993). Conversion of truncated and elongated prion proteins into the scrapie isoform in cultured cells. Proc Natl Acad Sci USA *90*, 3182-3186.

Sailer, A., Bueler, H., Fischer, M., Aguzzi, A., and Weissmann, C. (1994). No propagation of prions in mice devoid of PrP. Cell *77*, 967-968.

Sanghera, N., and Pinheiro, T.J. (2002). Binding of prion protein to lipid membranes and implications for prion conversion. J Mol Biol *315*, 1241-1256.

Silbereis, J., Cheng, E., Ganat, Y.M., Ment, L.R., and Vaccarino, F.M. (2009). Precursors with glial fibrillary acidic protein promoter activity transiently generate GABA interneurons in the postnatal cerebellum. Stem Cells *27*, 1152-1163.

Stahl, N., Borchelt, D.R., Hsiao, K., and Prusiner, S.B. (1987). scrapie prion protein contains a phosphatidylinositol glycolipid. Cell *51*, 229-240.

Taraboulos, A., Scott, M., Semenov, A., Avrahami, D., Laszlo, L., and Prusiner, S.B. (1995). Cholesterol depletion and modification of COOH-terminal targeting sequence of the prion protein inhibit formation of the scrapie isoform. J Cell Biol *129*, 121-132.

Tateishi, J., Kitamoto, T., Doh-ura, K., Boellaard, J.W., and Peiffer, J. (1992). Creutzfeldt-Jakob disease with amyloid angiopathy: diagnosis by immunological analyses and transmission experiments. Acta Neuropathol *83*, 559-563.

Trevitt, C.R., and Collinge, J. (2006). A systematic review of prion therapeutics in experimental models. Brain *129*, 2241-2265.

Trifilo, M.J., Yajima, T., Gu, Y., Dalton, N., Peterson, K.L., Race, R.E., Meade-White, K., Portis, J.L., Masliah, E., Knowlton, K.U., *et al.* (2006). Prion-induced amyloid heart disease with high blood infectivity in transgenic mice. Science *313*, 94-97.

Vey, M., Pilkuhn, S., Wille, H., Nixon, R., DeArmond, S.J., Smart, E.J., Anderson, R.G., Taraboulos, A., and Prusiner, S.B. (1996). Subcellular colocalization of the cellular and scrapie prion proteins in caveolae-like membranous domains. Proc Natl Acad Sci USA *93*, 14945-14949.

Walmsley, A.R., Zeng, F., and Hooper, N.M. (2003). The N-terminal region of the prion protein ectodomain contains a lipid raft targeting determinant. J Biol Chem *278*, 37241-37248.

Weissmann, C., Bueler, H., Fischer, M., Sauer, A., and Aguet, M. (1994). Susceptibility to scrapie in mice is dependent on PrP^C. Philos Trans R Soc Lond B Biol sci *343*, 431- 433.

White, M.D., Farmer, M., Mirabile, I., Brandner, S., Collinge, J., and Mallucci, G.R. (2008). Single treatment with RNAi against prion protein rescues early neuronal dysfunction and prolongs survival in mice with prion disease. Proc Natl Acad Sci USA *105*, 10238-10243.

LaVergne, TN USA
05 April 2010
1753LVUK00005B

* 9 7 8 0 9 5 4 3 3 3 5 2 2 *